Culinary Arts

PRINCIPLES AND AP

S

THIRD EDITION

STUD IDE

AMERICAN TECHNICAL PUBLISHERS
Orland Park, Illinois

Michael J. McGreal

The authors and publisher are grateful for the technical assistance provided by the following companies and organizations:

All-Clad Metalcrafters
American Lamb Board
American Metalcraft, Inc.
Barilla America, Inc.
The Beef Checkoff
Browne-Halco (NJ)
Bunn-O-Matic Corporation
Canada Beef, Inc.
Canada Cutlery Inc.
Carlisle FoodService Products
Chef's Choice® by EdgeCraft Corporation
Cooper-Atkins Corporation
Cres Cor
Detecto, A Division of Cardinal Scale Manufacturing Co.
Dexter-Russell, Inc.
Edlund Co.
Fluke Corporation
Frieda's Specialty Produce
Hobart
In-Sink-Erator
InterMetro Industries Corporation
Matfer Bourgeat USA
Melissa's Produce
Mercer Cutlery
Messermeister
National Oceanic and Atmospheric Administration
National Pork Board
National Turkey Federation
Paderno World Cuisine
Plitt Seafood
PolyScience
Shenandoah Growers
Strauss Free Raised
Tanimura & Antle®
Trails End Chestnuts
True FoodService Equipment, Inc.
United States Department of Agriculture
U.S. Fish and Wildlife Service
U.S. Geological Survey
U.S. Range
Vita-Mix® Corporation
The Vollrath Company, LLC
Vulcan-Hart, a Division of the ITW Food Equipment Group, LLC

3 4 5 6 7 8 9 – 18 – 9 8 7 6 5 4 3 2 1

Printed in the United States of America

ISBN 978-0-8269-4258-6

 This book is printed on recycled paper.

Contents

Chapter 1: **Foodservice Professionals**

Chapter 2: **Food Safety and Sanitation**

Chapter 3: **Knife Skills**

Chapter 4: **Tools and Equipment**

Chapter 5: **Cost Control Fundamentals**

Chapter 6: **Menu Planning and Nutrition**

Chapter 7: **Cooking Techniques**

CONTENTS

Chapter 12: **Fruits**

Chapter 13: **Vegetables**

Chapter 14: **Potatoes, Grains, and Pastas**

Chapter 15: **Garde Manger Fundamentals**

Contents

Chapter 16: **Poultry, Ratites, and Related Game**

Chapter 17: **Fish, Shellfish, and Related Game**

Chapter 18: **Beef, Veal, and Bison**

Chapter 19: **Pork and Related Game**

Chapter 20: **Lamb and Specialty Game**

Chapter 21: **Baking and Pastry Fundamentals**

INTRODUCTION

The *Culinary Arts Principles and Applications Study Guide* is designed to reinforce the concepts presented in *Culinary Arts Principles and Applications*. Each chapter in the study guide covers information from the corresponding chapter in the textbook by using review questions and activities to verify and reinforce comprehension of key concepts.

Review Questions consist of true-false, multiple choice, completion, and matching questions covering objectives, vocabulary terms, and important concepts that verify comprehension of content covered in each corresponding chapter of the textbook.

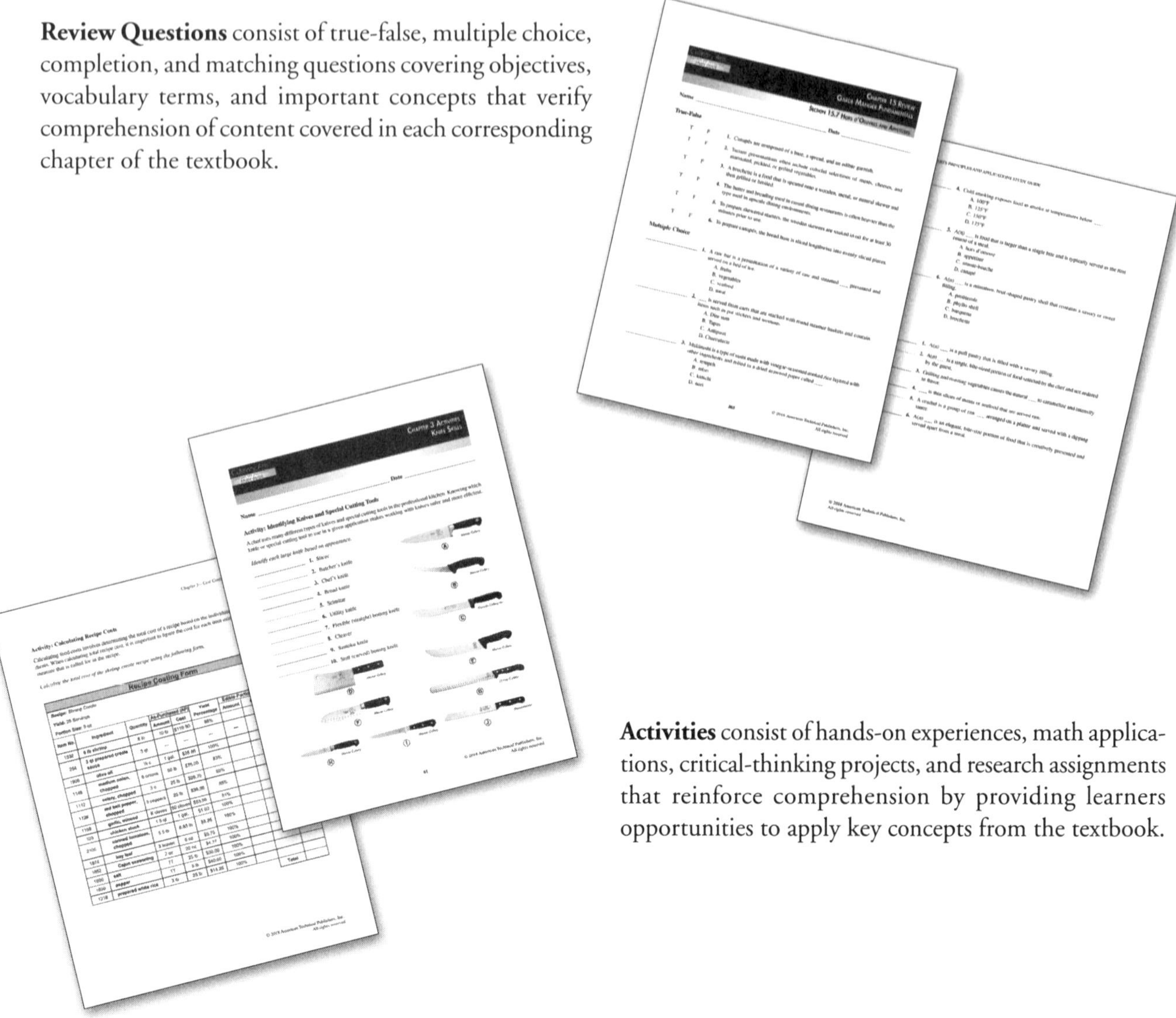

Activities consist of hands-on experiences, math applications, critical-thinking projects, and research assignments that reinforce comprehension by providing learners opportunities to apply key concepts from the textbook.

This study guide enables learners to review and apply knowledge as they study *Culinary Arts Principles and Applications*. It is an indispensable companion to the textbook that every learner will use. To obtain information about related training products, visit the American Tech website at www.atplearning.com.

The Publisher

Chapter 1 Review
Foodservice Professionals

Section 1.1 The Foodservice Industry

Name ______________________________ **Date** ____________________

True-False

T F **1.** In a quick-casual operation, the orders of guests are placed at a counter and then a runner brings the food to the table.

T F **2.** In the 18th century, guilds were commissioned by French royalty to prepare foods for royal celebrations.

T F **3.** Avant-garde cuisine emphasizes combining the flavors of two or more cultural regions.

T F **4.** Regardless of the foodservice operation, meal service style, or type of cuisine, service is the key to meeting and exceeding guest expectations.

T F **5.** In plated service, beverages and soups are presented from the left side of the guest with the left hand in a clockwise direction.

T F **6.** New American cuisine is a style of cuisine based on preparing foods that have been obtained directly from a local farm or grower.

Multiple Choice

______________ **1.** ___ service is a style of meal service in which food is served to guests seated at a table positioned against a wall with benches on either side.
A. Buffet
B. Banquet
C. Banquette
D. Booth

______________ **2.** A significant contribution Escoffier made to classical cuisine was the simplification of the many varieties of ___ presented by Carême.
A. soups
B. sauces
C. stocks
D. bases

____________________ **3.** Grande cuisine, also known as ___ cuisine, is a style of cuisine in which intricate food preparation methods and large, elaborate presentations are used.

A. fusion
B. nouvelle
C. haute
D. classical

____________________ **4.** A(n) ___ service provides food to a group of people at a remote site or special event.

A. catering
B. fine dining
C. institutional
D. cafeteria

____________________ **5.** ___ cuisine is a style of cuisine that uses a variety of cooking techniques to combine the flavors of two or more cultural regions.

A. Haute
B. Avant-garde
C. Fusion
D. New American

____________________ **6.** ___ service can include cooking foods tableside as guests observe.

A. French
B. English
C. American
D. Russian

____________________ **7.** The key to providing quality service is for staff to work together to provide guests with a ___ dining experience.

A. classical
B. positive
C. plated
D. grande

____________________ **8.** A(n) ___ customer is an individual who is the recipient of products or services within the same organization.

A. host
B. guest
C. external
D. internal

____________________ **9.** ___ service is a style of meal service in which all food is placed on the table for guests to pass for self-service.

A. Banquette
B. French
C. Family-style
D. Russian

______ **10.** A(n) ___ is a cart equipped with the items necessary to prepare foods tableside.
A. gueridon
B. banquette
C. restaurer
D. escoffier

Completion

______ **1.** ___ is a style of cuisine in which foods are cooked quickly, seasoned lightly, and artistically presented in smaller portions.

______ **2.** ___ service is a style of meal service in which food is displayed on a table that guests approach for self-service.

______ **3.** A(n) ___ is the way in which guests are served food and beverages.

______ **4.** ___ cuisine stresses the use of the finest ingredients and the most appropriate preparation methods to produce the best results but does not focus on presentation.

______ **5.** ___ service is a style of meal service in which a server holds a tray of food and serves guests food from the tray.

______ **6.** ___ service is commonly used at receptions or cocktail parties where bite-sized portions of food are offered.

______ **7.** Plated service, also known as ___ service, is a style of meal service in which individual portions of food are plated before being brought to guests.

______ **8.** ___ is a method of food production that seeks to conserve land and water, reduce energy consumption, prevent the use of harmful fertilizers and pesticides, and limit air pollution.

Matching

__________________ **1.** Grande cuisine

__________________ **2.** Classical cuisine

__________________ **3.** Fusion cuisine

__________________ **4.** Avant-garde cuisine

__________________ **5.** Nouvelle cuisine

__________________ **6.** New American cuisine

__________________ **7.** Farm-to-table cuisine

A. A style of cuisine in which intricate food preparation methods and large, elaborate presentations are used

B. A style of cuisine that uses a variety of cooking techniques to combine the flavors of two or more cultural regions

C. A style of cuisine in which foods are prepared using a formalized system of cooking techniques and are presented in courses

D. A style of cuisine, based on food chemistry, that involves the manipulation of the textures and temperatures of familiar dishes or ingredients to reinvent and present them in new and creative ways

E. A style of cuisine that emphasizes the use of foods that are grown in America

F. A style of cuisine in which foods are cooked quickly, seasoned lightly, and artistically presented in smaller portions

G. A style of cuisine based on preparing foods that have been obtained directly from a local farm or grower

Section 1.2 Foodservice Careers

Name ______________________________ **Date** ______________

True-False

T F **1.** A general manager is a person who directs an operation and oversees food production, sales, and service.

T F **2.** Food stylists express their creativity by arranging foods for photo shoots.

T F **3.** A sommelier is the person responsible for overseeing a specific production area of the kitchen.

T F **4.** The back-of-house includes the entry area, dining room, bar area, and public restrooms.

T F **5.** Servers have the most contact with guests and greatly influence the dining experience.

Multiple Choice

______________ **1.** The duties of a(n) ___ commonly include relaying dining room orders to various kitchen stations and reviewing each plate for accuracy and presentation before it leaves the kitchen.

A. station chef
B. sous chef
C. saucier
D. expediter

______________ **2.** ___ programs consist of both classroom instruction and hands-on application instruction with a chef-mentor.

A. On-the-job training
B. Apprenticeship
C. Certificate
D. Degree

______________ **3.** A ___ is a person responsible for serving alcoholic beverages from behind a bar.

A. bartender
B. server
C. maître d'
D. sommelier

______________ **4.** When the chef is absent, the ___ is in charge of the kitchen.
A. station chef
B. general manager
C. sous chef
D. porter

______________ **5.** A(n) ___ is earned by completing a food safety and sanitation program.
A. accreditation
B. degree
C. apprenticeship
D. certificate

Completion

______________ **1.** A(n) ___ and a general manager are examples of management positions commonly found throughout the foodservice industry.

______________ **2.** A(n) ___, also known as a dining room supervisor, is the person responsible for overseeing and coordinating all FOH activities.

______________ **3.** A(n) ___ is an individual enrolled in a formal training program who learns by practical experience under the supervision of a skilled professional.

______________ **4.** A(n) ___ is a person who operates the warewashing machine and cleans all of the pots, pans, dinnerware, glassware, and flatware.

______________ **5.** A(n) ___ is a person responsible for greeting and seating guests.

Matching

______________ **1.** Brigade system

______________ **2.** Apprentice

______________ **3.** Restaurateur

______________ **4.** Front-of-house

______________ **5.** Back-of-house

A. The portion of a foodservice operation that is typically not open to guests

B. An individual enrolled in a formal training program who learns by practical experience under the supervision of a skilled professional

C. A person who holds the legal title of a foodservice operation

D. A structured chain of command in which specific duties are aligned with the stations to which staff are assigned

E. The portion of a foodservice operation that includes the entry area, dining room, bar area, and public restrooms

Chapter 1 Review
Foodservice Professionals

Section 1.3 Employability Skills

Name ______________________________ **Date** ____________________

True-False

T F **1.** Reading skills are used to interpret menus, follow recipes, and check the accuracy of guest checks.

T F **2.** A positive attitude leads to increased productivity and results in the best products and services for guests.

T F **3.** If an employee exhibits ethics in the workplace, the employee does not have to be reliable, punctual, or prepared.

T F **4.** Effective communication skills are important for giving and understanding instructions and sharing information but not for providing feedback.

T F **5.** Recognizing and fulfilling a need such as refilling water glasses or straightening a work station before being asked does not show initiative.

T F **6.** Guests expect safe food in a clean environment offered by staff members who are well-groomed and healthy.

Multiple Choice

______________ **1.** When communicating in writing, it is important to ___ so that the message is not misunderstood.
- A. write legibly
- B. use permanent red ink
- C. underline specific requests
- D. capitalize entire words or sentences

______________ **2.** A speaking tone should always be professional, be respectful, and sound ___.
- A. condescending
- B. welcoming
- C. pointed
- D. close-minded

______________ **3.** Personal ___ practices such as bathing daily and following proper handwashing techniques are essential for all foodservice staff.
- A. protection
- B. safety
- C. hygiene
- D. communication

______________ **4.** Staff who look ___ make a positive first impression that reflects the quality of service guests can expect during their dining experience.
A. professional
B. stern
C. morose
D. speculative

______________ **5.** Instead of seeing obstacles, individuals with a positive attitude see ___.
A. unsolvable problems
B. challenges with solutions
C. closed doors
D. hurdles

______________ **6.** Staff should always speak in a clear and ___ tone that is loud enough to be heard.
A. rapid
B. pointed
C. pleasant
D. sing-song

Completion

______________ **1.** Effective communication requires speaking, ___, writing, and reading.

______________ **2.** Wearing appropriate attire, being dependable, and having a positive ___ are key employability skills.

______________ **3.** Purchase orders, memos, résumés, and job applications are forms of ___ communication used in foodservice operations.

______________ **4.** ___ among staff creates an environment that fosters mutual respect and collaboration.

______________ **5.** It is important to refrain from working when seriously ___ to prevent jeopardizing the health of others.

______________ **6.** Having a positive ___ in the workplace makes a staff member stand out.

Chapter 1 Review
Foodservice Professionals

Section 1.4 Front-of-House Fundamentals

Name ______________________________ **Date** ____________________

True-False

T F **1.** In order to fulfill guest expectations, FOH staff need to have menu knowledge.

T F **2.** There are many risks and liability concerns associated with serving alcohol, which can result in fines, loss of a liquor license, or imprisonment.

T F **3.** When making change for a guest, the first step is to return bills and then return any coins.

T F **4.** A POS system creates a guest check that documents what is sold to the guest.

T F **5.** Providing accommodations, such as crayons and paper, will cause a disruptive dining experience.

Multiple Choice

______________ **1.** According to the ___, businesses serving the public must make reasonable accommodations for guests with disabilities.
A. Food and Drug Administration
B. Americans with Disabilities Act
C. Environmental Protection Agency
D. Center for Disease Control

______________ **2.** ___ is the act of modifying something in response to a need or request.
A. An accommodation
B. Compliance
C. Expediting
D. Sidework

______________ **3.** Setting tables, stocking side areas, and rearranging furniture are tasks required from the ___.
A. FOH staff
B. expediter
C. general manager
D. bussers

_______________ **4.** ___ may consist of brewing a fresh batch of coffee or restocking glassware.
A. A point-of-sale-system
B. Sidework
C. Accommodations
D. Menu knowledge

_______________ **5.** If a guest uses a $20 bill to pay for a guest check with a total of $13.75, the guest is given ___ in change.
A. $5.25
B. $5.75
C. $6.00
D. $6.25

Completion

_______________ **1.** ___ staff are responsible for preparing the dining room before guests arrive.

_______________ **2.** ___ is the process of speeding up the ordering, preparation, and delivery of food to guests.

_______________ **3.** ___ is the process of cleaning, restocking, and preparing the items needed to keep service running smoothly.

_______________ **4.** A guest ___ may include the amount due for food and beverages, sales tax, and service charges.

_______________ **5.** A(n) ___ system is a computerized network that compiles data on all sales incurred by a business.

Matching

_______________ **1.** Accommodation

_______________ **2.** Front-of-house (FOH) staff

_______________ **3.** Point-of-sale (POS) system

_______________ **4.** Expediting

_______________ **5.** Sidework

A. The process of speeding up the ordering, preparation, and delivery of food to guests

B. The act of modifying something in response to a need or request

C. The process of cleaning, restocking, and preparing the items needed to keep service running smoothly

D. A computerized network that compiles data on all sales incurred by a business

E. The staff who are responsible for organizing and assembling items necessary for dining room service

Chapter 1 Review
Foodservice Professionals

Section 1.5 Back-of-House Fundamentals

Name ______________________________ **Date** ____________________

True-False

T F **1.** Mise en place makes efficient use of time and keeps preparation areas organized.

T F **2.** Adjusting a recipe requires calculating new amounts for only dry ingredients.

T F **3.** New discoveries about food make it essential to continuously increase food knowledge.

T F **4.** Accommodations are often made by BOH staff for guests with food allergies and for individuals following special diets.

T F **5.** FOH and BOH skill sets are not important to the success of a foodservice operation.

Multiple Choice

______________ **1.** ___ must know how to execute cooking methods and how to properly portion and plate food.
- A. BOH staff
- B. FOH staff
- C. Hosts
- D. Servers

______________ **2.** Mise en place is a French term meaning "___."
- A. plated dish
- B. put in place
- C. proper measurement
- D. presentation ready dish

______________ **3.** To keep food production and service efficient, BOH staff must effectively ___ FOH staff.
- A. close communication with
- B. switch places with
- C. ignore
- D. interact with

Completion

______________ **1.** A food ___ may limit what an individual eats and can be life threatening.

______________ **2.** ___ knowledge includes knowing which cooking methods and ingredients should be used to effectively prepare and enhance the flavor of each dish.

______________ **3.** Knowing the units of ___ and the difference between weight and volume are vital to accurately measuring ingredients.

______________ **4.** BOH staff ___ dishes by checking to make sure food is correctly portioned, plated correctly, and the proper temperature.

______________ **5.** Providing guests with a list of ingredients used to prepare dishes, being able to substitute or eliminate specific ingredients, and preparing food in a special area of the kitchen are ways BOH staff ___ guests with food allergies.

Chapter 1 Review
Foodservice Professionals

Section 1.6 Employment Preparation

Name ______________________________ Date ____________________

True-False

T F 1. Questions should not be asked about the foodservice operation during a job interview.

T F 2. A job interview helps reveal the communication skills, confidence, and integrity of the applicant.

T F 3. A professional résumé can make a positive impression and show an employer the applicant is prepared and organized.

T F 4. Someone who has no experience in a restaurant and is looking for a first job may have the following career objective: To acquire a position in a quality foodservice operation in order to gain real-life work experience.

Multiple Choice

__________ 1. Networking is a means of using ___ to locate possible employment opportunities.
- A. help wanted signs
- B. personal connections
- C. printed resources
- D. public advertisements

__________ 2. A ___ should list the applicant's level of education, date of graduation, school or college name, and diploma or certificate title.
- A. portfolio
- B. job listing
- C. résumé
- D. career objective

__________ 3. ___ is(are) the most common means of finding employment.
- A. Networking
- B. Online job postings
- C. Printed resources
- D. Commercial advertising

__________ 4. If an applicant is looking for a pastry job, a ___ would likely include images of pastries the applicant has made.
- A. job posting
- B. job application
- C. résumé
- D. portfolio

Completion

__________________ **1.** A(n) ___ is a direct statement expressing the type of employment goal being sought.

__________________ **2.** A(n) ___ is a document listing the education, professional experience, and interests of a job applicant.

__________________ **3.** A(n) ___ is a collection of items that depict the knowledge, skills, and accomplishments of an individual.

__________________ **4.** An employment ___ requires applicants to list basic information such as name, address, relevant work experience, educational background, position desired, and availability.

__________________ **5.** After an interview, a(n) ___ letter is sent to the interviewer.

Matching

__________________ **1.** Résumé

__________________ **2.** Portfolio

__________________ **3.** Carrer objective

__________________ **4.** Networking

A. A document listing the education, professional experience, and interests of a job applicant

B. A direct statement expressing the type of employment goal being sought

C. A means of using personal connections, such as contacts through professional organizations, friends, teachers, and acquaintances, to locate possible employment opportunities

D. A collection of items that depict the knowledge, skills, and accomplishments of an individual

Name ______________________________ **Date** ____________________

Activity: Researching the Foodservice Industry

Delivering quality food and service is vital for foodservice operations. If the service is poor, guests may choose not to return to a foodservice operation even if the food is memorable. Likewise, if the service is superior, guests are less likely to return for a second meal if the food is poor. The key to success in the foodservice industry involves a staff committed to working together to provide guests with a positive dining experience.

Research an existing foodservice establishment. Give a complete overview of the operation using the following outline.

I. Introduction
II. Type of restaurant
III. History of the company
  A. Who founded the company?
  B. Date company was founded
  C. Where the company was founded
IV. Current operations
  A. Menu
  B. Financial data (if available)
V. Reasons for failure/success
VI. Conclusion

1. Using the outline, prepare a report and present findings to the class. Use visuals, such as maps, pictures, a historical timeline, food demonstrations, etc., to enhance the report. Include a list of sources used for the project.

2. Write a short summary of what was learned from this assignment.

Activity: Analyzing Meal Service Styles

There are many factors that contribute to the success of a foodservice operation. Location often influences the type of dining operation that can be successful in a given area. However, a given area may not be able to support too many of the same type of dining operations. The type of dining operation often dictates the meal service style provided by the staff. Common meal service styles include booth, banquette, buffet, plated, English, Russian, French, butler, and banquet service.

Research 10 local foodservice operations and use that information to complete the chart below and the questions that follow. Several operations may feature one particular meal service style, and one or more meal service styles may be missing from the area.

Restaurant Name	Service Style
1.	
2.	
3.	
4.	
5.	
6.	
7.	
8.	
9.	
10.	

11. Based on the data gathered, are there any meal service styles that are more popular than others in the area?

12. If yes, why are these meal service styles more popular?

13. Which meal service styles are missing in the area?

14. Which missing meal service styles, if any, might also be successful?

15. How does a dining environment influence the meal service style?

16. What conclusions can be drawn from this data collection?

Activity: Researching Influential Chefs

Throughout the years, various types of cuisine have evolved. The evolution of cuisine is due to the creativity of individual chefs and the advances in technology, food products, and preparation techniques, which have led to inspired and innovated dishes.

Research a pioneer of the foodservice industry or a contemporary food service professional. Use the following outline to give a complete overview of the chef's life and accomplishments.

I. Introduction
II. Country of origin
III. Date of birth
IV. Education and training
V. History of work in foodservice
VI. Major accomplishments
  A. Innovative procedures
  B. Famous dishes/recipes
VII. Effects of accomplishments on contemporary culinary arts
VIII. Conclusion

1. Using the outline, prepare a report and present the findings to the class. Use visuals, such as maps, pictures, a historical timeline, food demonstrations, etc., to enhance the report. Include a list of sources used.

2. Write a short summary of what was learned from this assignment.

Activity: Making Change

When a customer pays with cash, it is often necessary to return change. The amount of change given to the customer should be equal to the difference between the amount of cash presented for payment and the total amount of the guest check. Instead of simply handing the customer the change all at once, it is appropriate to return it to the customer in an orderly process. This reduces the chance of returning the incorrect amount of change.

Complete the following statements using the guest check below. The guest pays for the check with a $20.00 bill.

____________________ **1.** The total amount due is $____.

____________________ **2.** The first amount of change (coin) that should be returned is $____.

____________________ **3.** The second amount of change (small bill) that should be returned is $____.

____________________ **4.** The third amount of change (large bill) that should be returned is $____.

GUEST CHECK

Date	Table	Guests	Server	47221
1/5	10	1	#123	

1	Sandwich	7	99
1	Cup of Soup	3	75
1	Iced Tea		99
	Subtotal	12	73
	Discount		
	Sales Tax	1	02
	Total	13	75

A 15% gratuity will be added to groups of 8 or more.

Activity: Identifying FOH and BOH Skill Sets

Both FOH (front-of-house) and BOH (back-of-house) skill sets are equally important to the success of a foodservice operation. When these skills are combined effectively, collaboration and teamwork result in a positive dining experience for guests.

Identify the following FOH, BOH, or both FOH and BOH tasks.

_______________ **1.** Measuring ingredients

_______________ **2.** Making menu recommendations

_______________ **3.** Filling salt and pepper shakers

_______________ **4.** Chopping lettuce

_______________ **5.** Rearranging furniture to accommodate guests

_______________ **6.** Maintaining good personal hygiene

_______________ **7.** Portioning entrées onto plates

_______________ **8.** Wearing professional work attire

_______________ **9.** Setting tables

_______________ **10.** Cooking foods properly

_______________ **11.** Using appropriate cooking equipment

_______________ **12.** Making change for guests

_______________ **13.** Exhibiting a positive attitude

_______________ **14.** Communicating effectively

_______________ **15.** Adjusting recipes

_______________ **16.** Being reliable, punctual, and prepared

_______________ **17.** Using knives properly

_______________ **18.** Reading food orders out loud to cooks

_______________ **19.** Preparing food with alternative ingredients

_______________ **20.** Providing a booster chair for a child

Activity: Determining Mise en Place

Mise en place is a French term meaning "put in place." Mise en place makes efficient use of time and keeps preparation areas organized. To help ensure food preparation is carried out in an orderly and efficient manner, it is essential for BOH staff to complete the necessary mise en place prior to receiving the orders from guests.

Identify components from the following recipe that could be prepared ahead of time as the mise en place.

Triple-Decker Turkey Club Sandwiches

6 slices sandwich bread

3 tbsp mayonnaise

6 romaine lettuce leaves

1 medium tomato, sliced

4 oz turkey breast, cooked and sliced

4 slices bacon, cooked and cut into 4 inch lengths

1. How can mise en place be applied to the romaine lettuce leaves?

2. How can mise en place be applied to the tomato?

3. How can mise en place be applied to the turkey breast?

4. How can mise en place be applied to the bacon?

Activity: Developing Portfolios

A portfolio is a collection of items depicting a job candidate's talents, skills, accomplishments, and overall abilities and is used to market the job candidate to a potential employer for a particular career. A portfolio should be updated continually. Information regarding job experience should also be included. Special attention should be given to information that reflects career goals. Letters of recommendation from an instructor, a counselor, and/or employers should be included.

Present information in a professional-looking final form such as a notebook, binder, photo album, or scrapbook. The presentation should be neat and organized. Information in a portfolio is often organized into the following categories:

- Cover letter
- Résumé
- Awards and certificates
- Photos of work
- Letters of reference

1. Prepare a sample portfolio using your job experience and skills. Include the items listed in the description above.

2. Submit a portfolio in final form. Information should be typed, well-organized, and professional.

Activity: Analyzing the Hiring Process

A manager is responsible for maintaining a full staff at a foodservice operation. The manager must identify skills that are necessary to perform the job and the qualities that are most desirable in a potential employee. Any skills, qualities, education, and work experience required for the job are included in a job description. The job description also summarizes duties and responsibilities of the position. Job applicants are evaluated against criteria in the job description during an interview.

Prepare a job description for the open position of a sous chef, line cook, maître d', host, server, or busser by answering the following questions.

1. What is the title of the position?

2. What is the name of the establishment?

3. Where is the establishment located?

4. List four responsibilities of the position.

5. List four qualities that are required for the position.

6. What education or training is required for the position?

7. What work experience is required for the position?

8. What skills or physical requirements are needed to perform the duties of the position?

Prepare six open-ended questions (questions that cannot be answered with a simple yes or no response) to use in an interview for the position described in items 1–8. Then, interview a classmate for that position and record their answers. Observe the nonverbal communication that is displayed during the interview and note it along with the candidate's verbal responses.

9. What is the first interview question?

10. What was the candidate's response to question 1?

11. What is the second interview question?

12. What was the candidate's response to question 2?

13. What is the third interview question?

14. What was the candidate's response to question 3?

15. What is the fourth interview question?

16. What was the candidate's response to question 4?

17. What is the fifth interview question?

18. What was the candidate's response to question 5?

19. What is the sixth interview question?

20. What was the candidate's response to question 6?

21. Evaluate the experience and take note of what might be done differently to make the process more successful. Submit findings in a written report. Include the job description, interview questions, and evaluation.

Chapter 2 Review
Food Safety and Sanitation

Section 2.1 Food Safety

Name ______________________________ Date ______________

True-False

T F 1. Entities that enforce food safety codes have the authority to shut down foodservice operations that fail to meet local safety codes.

T F 2. Some states and employers have even more stringent guidelines than ServSafe® regarding certification renewal.

Multiple Choice

______________ 1. ServSafe® certification is valid for ___ years.
- A. two
- B. three
- C. five
- D. six

______________ 2. ___ is a government agency that ensures protection of animals and plants from diseases or pests.
- A. NSF International®
- B. U.S. Food and Drug Administration
- C. USDA Food Safety Inspection Service
- D. USDA Animal and Plant Health Inspection Service

Completion

______________ 1. The ___ is a government agency that investigates foodborne illnesses and plays a key role in supporting state and local health departments.

______________ 2. Food safety standards are enforced by local ___ departments and other authorities having jurisdiction.

Chapter 2 Review
Food Safety and Sanitation

Section 2.2 Food Contamination

Name ______________________________ **Date** ____________________

True-False

T F **1.** Chemical contaminants include cleaning supplies and pesticides.

T F **2.** A virus is a pathogen that grows inside the cells of a host.

T F **3.** A biological contaminant is a living microorganism that can be hazardous if inhaled, swallowed, or otherwise absorbed.

T F **4.** Bacteria thrive on protein-rich foods for the energy they need to reproduce.

T F **5.** Cross-contamination is contamination that occurs when food is exposed to the original source of the contaminant.

Multiple Choice

______________ **1.** The temperature danger zone is a range of temperature, between 41°F and ___, in which bacteria thrive.
A. 125°F
B. 130°F
C. 135°F
D. 140°F

______________ **2.** Almost all foodborne pathogens require ___ to grow.
A. oxygen
B. minerals
C. mold
D. acidity

______________ **3.** Foods such as poultry, seafood, meats, and soft cheeses contain high amounts of ___ and are therefore more susceptible to fast-growing bacteria.
A. sugar
B. vitamins
C. oxygen
D. water

______ **4.** Every food falls between 0 and ___ on the pH scale.
A. 7
B. 10
C. 14
D. 16

______ **5.** ___ are pathogens that live in soil, water, or organic matter.
A. Parasites
B. Viruses
C. Bacteria
D. Molds

Completion

______ **1.** ___ occurs when food is exposed to harmful elements through an intermediate, or indirect, carrier.

______ **2.** ___ is the acronym that identifies the six factors that contribute to bacterial growth.

______ **3.** ___ multiply on acidic foods that have low water content, such as breads.

______ **4.** A parasite is a pathogen that relies on a(n) ___ for survival in a way that benefits only the organism.

______ **5.** Most bacteria levels are safe at temperatures of ___°F or above.

Matching

______ **1.** Foodborne illness

______ **2.** Physical contaminant

______ **3.** Contaminated food

______ **4.** Chemical contaminant

______ **5.** Food spoilage indicator

______ **6.** Biological contaminant

A. Any object that can be hazardous if it is inhaled, swallowed, or otherwise absorbed into the body

B. Any chemical substance that can be hazardous if it is inhaled, swallowed, or otherwise absorbed into the body

C. A living microorganism that can be hazardous if it is inhaled, swallowed, or otherwise absorbed

D. A condition that signifies that food is deteriorating and is no longer safe for consumption

E. An illness that is carried or transmitted to people through contact with or consumption of contaminated food

F. Food that is unsafe to eat due to the presence of harmful microorganisms, toxic substances, or foreign objects

Chapter 2 Review
Food Safety and Sanitation

Section 2.3 Sanitation

Name ______________________________ **Date** ____________________

True-False

T F **1.** Sanitizing is the process of removing food and residue from a surface.

T F **2.** Handwashing does not need to occur after disposing of garbage.

T F **3.** To eliminate and prevent infestations, incoming supplies should be checked for pests and garbage should not be allowed to accumulate with excess.

T F **4.** Washrooms used by foodservice employees must display signs notifying employees that they must wash their hands before returning to work.

T F **5.** Gloves are a substitute for handwashing.

Multiple Choice

______________ **1.** Heat sanitizing is the process of using very hot ___ to reduce harmful pathogens to a safe level.
- A. soap
- B. chemicals
- C. air
- D. water

______________ **2.** When handwashing, hands should be scrubbed vigorously for at least ___ seconds.
- A. 10
- B. 20
- C. 25
- D. 30

______________ **3.** Gloves should never be worn for more than ___ hours.
- A. 2
- B. 3
- C. 4
- D. 5

_______________ **4.** All food contact surfaces must be cleaned and sanitized after each use, before working with any food, or a minimum of every ___ hours.

A. 1½

B. 3

C. 4

D. 6

_______________ **5.** Heat sanitizing is immersing an object in water that is at least 171°F for a minimum of ___.

A. 30 seconds

B. 2 minutes

C. 5 minutes

D. 10 minutes

Completion

_______________ **1.** A(n) ___ is often used to wash and sanitize large items and items that are difficult to clean due to foods that have adhered to the items.

_______________ **2.** Handwashing stations must be located in ___, service, and dishwashing areas.

_______________ **3.** Except for a plain ___, do not wear jewelry.

_______________ **4.** ___ hands immediately after using a handkerchief or coughing.

_______________ **5.** The procedure for handwashing includes wetting hands and arms with hot water that is at least ___°F.

_______________ **6.** In a low-temperature dish machine, the chemical sanitizer needs to reach at least ___°F.

Matching

_______________ **1.** Sanitizing

_______________ **2.** Personal hygiene

_______________ **3.** Warewashing

_______________ **4.** Cleaning

A. The physical care maintained by an individual

B. The process of cleaning and sanitizing all items used to prepare and serve food

C. The process of destroying or reducing harmful microorganisms to a safe level

D. The process of removing food and residue from a surface

Chapter 2 Review
Food Safety and Sanitation

Section 2.4 The Food Path

Name ______________________________ Date ____________________

True-False

T F 1. Dry storage facilities should be kept dry, clean, and between 50°F and 70°F.

T F 2. Once food is received, inspections are not necessary to uphold quality and safety standards.

T F 3. When cooling a hot food, it must be cooled to 70°F within 2 hours.

T F 4. In top-to-bottom refrigerated storage, cooked and ready-to-use items are stored on the bottom shelf of the refrigerator.

T F 5. A potentially hazardous food is a food that requires temperature control in order to keep it safe for consumption.

Multiple Choice

______________ 1. A thermometer inserted into the ___ part of a food is the best way to determine its temperature.

A. hardest
B. softest
C. thickest
D. thinnest

______________ 2. Potentially hazardous foods must be maintained at a temperature at or below 41°F, with the exception of ___.

A. poultry
B. milk
C. meats
D. eggs in the shell

______________ 3. Freezers should be set at ___°F to keep all products frozen.

A. −41
B. −10
C. 28
D. 32

______________ **4.** All dry storage items must be kept at a minimum of ___ inches above the floor.
A. 6
B. 10
C. 12
D. 15

______________ **5.** Food can be thawed by submerging it under clean, running water that is ___°F or below.
A. 70
B. 72
C. 74
D. 76

Completion

______________ **1.** ___ is the process of dating new items as they are placed into inventory and placing them behind or below older items to ensure that older items are used first.

______________ **2.** When thawing food, no portion should reach ___°F or above.

______________ **3.** Whether a food is served hot or cold, its temperature needs to stay out of the ___ while it is being held for service.

______________ **4.** Transport containers must be well-insulated so that cold foods are held at 41°F or below and hot foods are held at ___°F or above.

______________ **5.** When cooling hot foods, use a cold ___ to cool foods such as soups, sauces, and stews.

Matching

______________ **1.** Potentially hazardous food

______________ **2.** FIFO

______________ **3.** Flow of food

A. The process of dating new items as they are placed into inventory and placing them behind or below older items to ensure that older items are used first

B. The path food takes in a foodservice operation as it moves from purchasing to service

C. A food that requires temperature control in order to keep it safe for consumption

Chapter 2 Review
Food Safety and Sanitation

Section 2.5 Food Safety Management

Name ______________________________ **Date** ______________

True-False

T F **1.** A critical limit is the point where a hazard can be prevented, eliminated, or reduced.

T F **2.** A HACCP plan must identify the individual responsible for monitoring food.

Multiple Choice

_______________ **1.** ___ is a form of technology that uses electronic tags to store data that can be monitored from remote distances.
A. Time monitoring
B. Distance monitoring
C. Electronic identification
D. Radio frequency identification

_______________ **2.** Logs, graphs, charts, receipts, and notes are examples of ___ that should be kept and maintained in order for a HACCP plan to be the most effective.
A. documentation
B. critical limits
C. certificates
D. process-specific lists

_______________ **3.** The ___ system is a food safety management system that aims to identify, evaluate, and control contamination hazards throughout the flow of food.
A. Food Code
B. HACCP
C. FDA
D. NSF

_______________ **4.** The FDA ___ establishes the implementation of a HACCP plan as a voluntary effort for foodservice operations.
A. Food Code
B. critical control point (CCP)
C. hazard analysis
D. corrective action

Completion

______________ **1.** A(n) ___ is the point in a HACCP plan that identifies the steps that must be taken when food does not meet a critical limit.

______________ **2.** A(n) ___ is the process of assessing potential risks in the flow of food in order to establish what must be addressed in the HACCP plan.

______________ **3.** A critical limit needs to be established for each ___.

______________ **4.** A(n) ___ is a written document detailing what policies and procedures will be followed to help ensure the safety of food.

SECTION 2.6 WORKPLACE SAFETY STANDARDS

Name ______________________________ Date ______________

True-False

T F 1. Proper lifting prevents back strain by using the legs to power the lift.

T F 2. Class A, Class D, and Class K fire extinguishers are commonly used in foodservice operations.

T F 3. The USDA is responsible for setting and enforcing workplace safety standards in the United States.

T F 4. Foodservice fire-suppression systems extinguish fires by spraying special chemicals from overhead nozzles when the system is activated.

T F 5. Second-degree burns affect only the top layer of skin and appear red and swollen.

Multiple Choice

______________ 1. When an accident occurs, a supervisor should be notified ___.
A. immediately
B. within the hour
C. within the day
D. within the week

______________ 2. A(n) ___ is a document that provides detailed information describing a chemical, instructions for its safe use, the potential hazards, and appropriate first-aid measures.
A. incident report
B. fire-suppression system
C. safety data sheet
D. emergency action plan

______________ 3. Each foodservice operation facility must have at least ___ properly marked fire exits that are clear of obstructions.
A. two
B. three
C. four
D. five

______________ **4.** Class ___ extinguishers are used for grease fires.
A. A
B. C
C. K
D. M

______________ **5.** A(n) ___ is a protective sleeve placed over the finger to prevent contamination of a cut.
A. finger cot
B. hazard tape
C. ace bandage
D. gauze

______________ **6.** Blisters, intense pain, and swelling result from ___ burns.
A. first-degree
B. second-degree
C. third-degree
D. minor

Completion

______________ **1.** OSHA standards require employers to provide a fire protection plan, fire exits, fire extinguishers, a fire-suppression system, and a(n) ___.

______________ **2.** A(n) ___ is a chemical present in the workplace that is capable of causing harm.

______________ **3.** A(n) ___ is a document that provides detailed information describing a chemical, instructions for its safe use, the potential hazards, and appropriate first-aid measures.

______________ **4.** A fire-suppression system is an automatic fire extinguishing system that is activated by the intense ___ generated by a fire.

______________ **5.** A(n) ___ is a written plan intended to organize employees during an emergency situation.

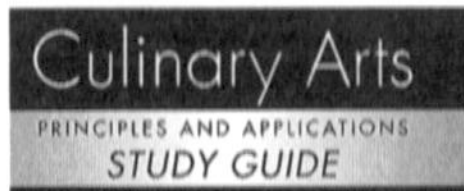

Name ______________________________ **Date** ____________________

Activity: Identifying Food Safety and Sanitation Acronyms

Acronyms are often used for safety and sanitation terms within the foodservice industry. Knowing these acronyms helps promote effective communication.

Write what each of the following acronyms represents.

1. What does the acronym HACCP stand for?
2. What does the acronym FDA stand for?
3. What does the acronym CDC stand for?
4. What does the acronym OSHA stand for?
5. What does the acronym CCP stand for?
6. What does the acronym FSIS stand for?
7. What does the acronym FAT TOM stand for?
8. What does the acronym UL stand for?
9. What does the acronym FIFO stand for?

Activity: Matching Food Safety and Sanitation Standards

Food safety and sanitation are vital to successful foodservice operations. It is the duty of every foodservice employee to follow safety and sanitation standards in both the kitchen and dining room.

Match the appropriate actions to each of the following numbers.

__________ **1.** Between 41°F and 135°F

__________ **2.** Below 4.6

__________ **3.** 10–15 seconds

__________ **4.** 70°F

__________ **5.** 160°F

__________ **6.** 145°F

__________ **7.** 4 hours

__________ **8.** 1–4 hours

__________ **9.** 41°F

__________ **10.** 2–4 hours

__________ **11.** 100°F

__________ **12.** 165°F

__________ **13.** 171°F

__________ **14.** 180°F

__________ **15.** 110°F

A. Minimum length of time to thoroughly scrub hands during handwashing

B. Water temperature required for a high-temperature dish machine

C. Minimum required temperature for heat sanitizing

D. Minimum internal cooking temperature for ground beef, lamb, and veal

E. The pH level that inhibits bacteria growth

F. Minimum internal temperature for beef, veal, and lamb steaks, chops, and roasts

G. Maximum water temperature when thawing foods under clean, running water

H. Minimum required water temperature in the first sink compartment for warewashing

I. Minimum internal cooking temperature for poultry

J. The range of temperature known as the temperature danger zone in which bacteria thrive

K. Maximum receiving temperature for ready-to-eat foods

L. The duration of the lag phase

M. Minimum required water temperature for handwashing

N. Time interval for checking the temperature of held foods

O. After a hot food is cooled to 70°F within 2 hours, the length of time the food has to reach 41°F

Activity: Identifying the Flow of Food

The flow of food is the path food takes in a foodservice operation as it moves from purchasing to service. Contaminants can enter at any stage in the flow of food. Understanding the flow of food is essential to prevent foodborne illness.

Complete the following chart to show the flow of food using the words below.

- Cook
- Preparing
- Reheat
- Receiving
- Cool

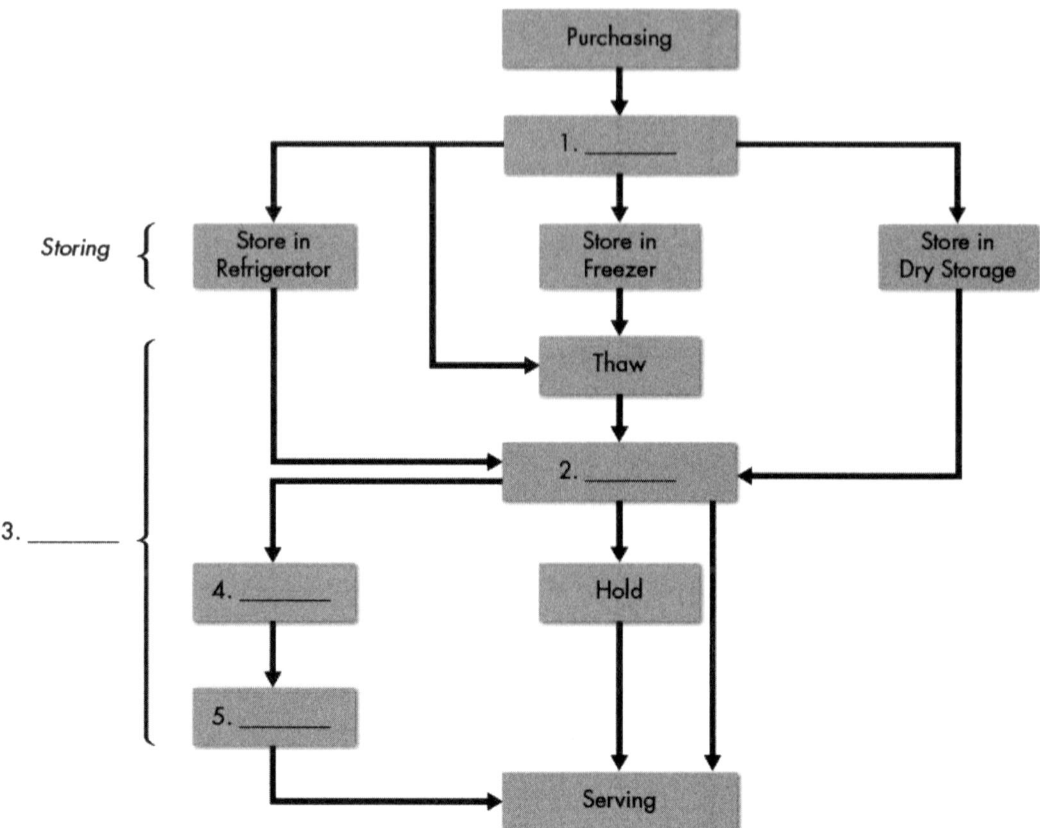

Activity: Matching HACCP Principles

The Hazard Analysis Critical Control Point (HACCP) system is a food safety management system that aims to identify, evaluate, and control contamination hazards throughout the flow of food. A HACCP system revolves around seven basic principles that can help ensure that food remains safe from the time it is received through the time of service.

Match the seven HACCP principles to the correct definition.

____________________	**1.** Corrective action	____________________	**5.** Monitoring procedure
____________________	**2.** Hazard analysis	____________________	**6.** System verification
____________________	**3.** Documentation	____________________	**7.** Critical control point
____________________	**4.** Critical limit	____________________	**8.** HACCP plan

A. Forms such as logs, graphs, charts, receipts, and notes that should be kept and maintained

B. The point in the HACCP plan that identifies the individual responsible for ensuring the safety of food

C. The process of assessing potential risks in the flow of food in order to establish what must be addressed in the HACCP plan

D. The point at which a hazard can be prevented, eliminated, or reduced

E. The point in a HACCP plan at which a standard such as a minimum or maximum value is established for a CCP in order to prevent, eliminate, or reduce a hazard to a safe level

F. The point in a HACCP plan that identifies the steps that must be taken when food does not meet a critical limit

G. The process of determining if the HACCP plan is preventing, reducing, or eliminating identified hazards

H. A written document that details the policies and procedures that will be followed to help ensure the safety of food

Activity: Arranging HACCP Principles

A HACCP plan is a written document detailing what policies and procedures will be followed to help ensure the safety of food. Although a HACCP plan is different for each foodservice operation, the seven HACCP principles must be followed in sequential order.

Place the seven basic HACCP principles in the order they must be followed.

____________________	**1.**	**A.** Identify corrective actions
____________________	**2.**	**B.** Establish critical limits
____________________	**3.**	**C.** Verify the system
____________________	**4.**	**D.** Conduct a hazard analysis
____________________	**5.**	**E.** Maintain documentation
____________________	**6.**	**F.** Determine critical control points
____________________	**7.**	**G.** Set monitoring procedures

Activity: Analyzing Safety Data Sheets

OSHA requires employers to tell employees about potential dangers of chemicals used in the workplace. A safety data sheet (SDS) contains valuable information that is needed in case of an emergency. The format used can vary for each operation but the following information should be included:

- name, address, and phone number of the manufacturer
- date the information was prepared, validated, or printed
- information about safe use, handling, and storage of the product
- physical, health, fire, and reactivity hazards
- first-aid information and steps to take in an emergency
- hazardous components
- appropriate personal protective equipment to wear when exposed to the chemical

Use the J-512 Sanitizer SDS to answer the questions and statements that follow.

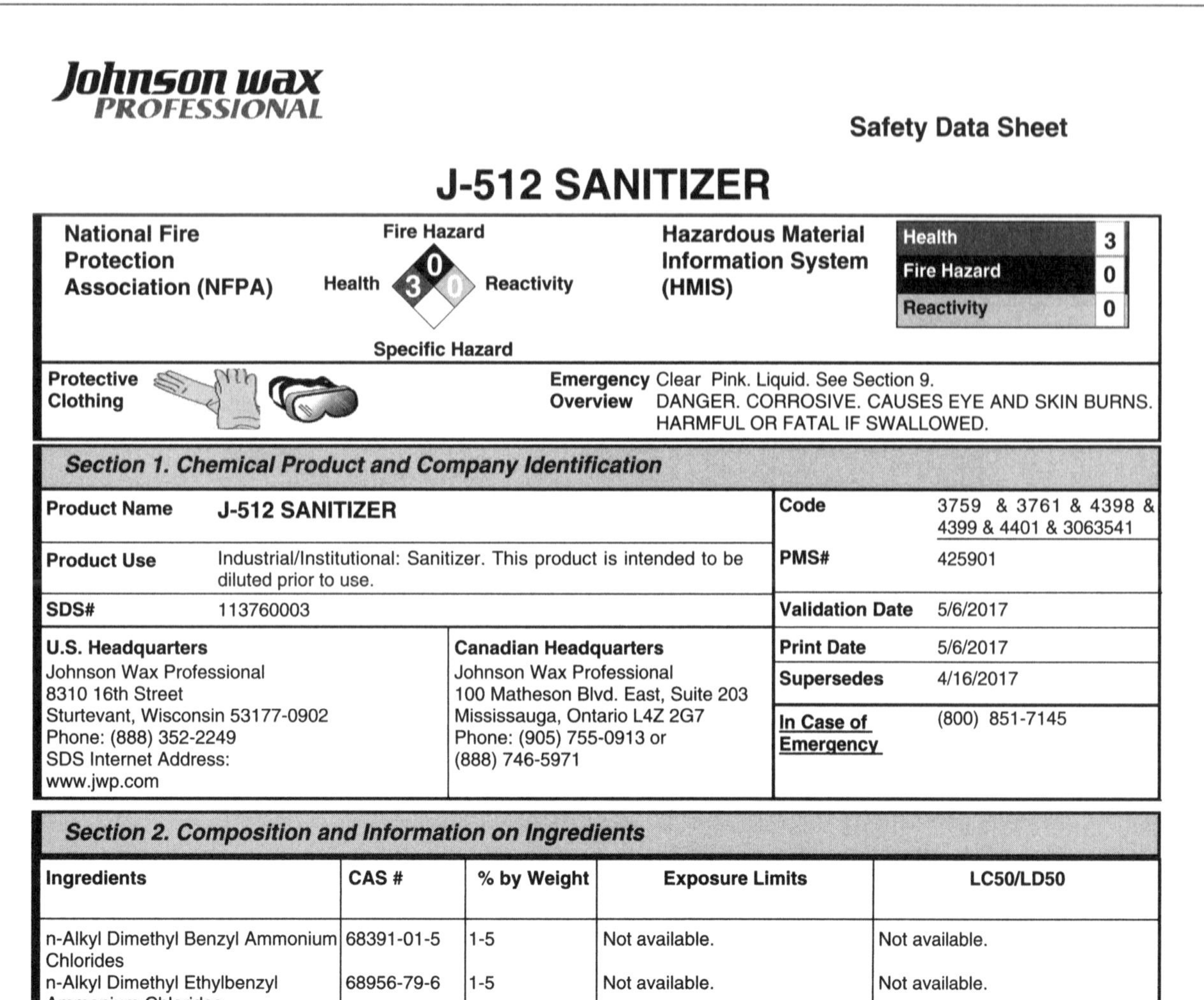

Johnson wax PROFESSIONAL

Safety Data Sheet

J-512 SANITIZER

National Fire Protection Association (NFPA)	Fire Hazard 0; Health 3; Reactivity 0; Specific Hazard	Hazardous Material Information System (HMIS)	Health 3; Fire Hazard 0; Reactivity 0

Protective Clothing		Emergency Overview	Clear Pink. Liquid. See Section 9. DANGER. CORROSIVE. CAUSES EYE AND SKIN BURNS. HARMFUL OR FATAL IF SWALLOWED.

Section 1. Chemical Product and Company Identification

		Code	3759 & 3761 & 4398 & 4399 & 4401 & 3063541
Product Name	J-512 SANITIZER		
Product Use	Industrial/Institutional: Sanitizer. This product is intended to be diluted prior to use.	PMS#	425901
SDS#	113760003	Validation Date	5/6/2017
		Print Date	5/6/2017
		Supersedes	4/16/2017
		In Case of Emergency	(800) 851-7145

U.S. Headquarters
Johnson Wax Professional
8310 16th Street
Sturtevant, Wisconsin 53177-0902
Phone: (888) 352-2249
SDS Internet Address:
www.jwp.com

Canadian Headquarters
Johnson Wax Professional
100 Matheson Blvd. East, Suite 203
Mississauga, Ontario L4Z 2G7
Phone: (905) 755-0913 or
(888) 746-5971

Section 2. Composition and Information on Ingredients

Ingredients	CAS #	% by Weight	Exposure Limits	LC50/LD50
n-Alkyl Dimethyl Benzyl Ammonium Chlorides	68391-01-5	1-5	Not available.	Not available.
n-Alkyl Dimethyl Ethylbenzyl Ammonium Chlorides	68956-79-6	1-5	Not available.	Not available.

Continued on Next Page

Page: 1/4

Safety Data Sheet

J-512 SANITIZER

Section 3. Hazards Identification

Routes of Entry	Inhalation. Skin contact. Eye contact.
Potential Acute Health Effects	
Eyes	Corrosive. May cause permanent damage including blindness.
Skin	Corrosive. May cause permanent damage.
Inhalation	May cause irritation and corrosive effects to nose, throat and respiratory tract.
Ingestion	Corrosive. May cause burns to mouth, throat, and stomach.
Medical Conditions Aggravated by Overexposure:	Individuals with chronic respiratory disorders such as asthma, chronic bronchitis, emphysema, etc., may be more susceptible to irritating effects.
See Toxicological Information (section 11)	

Section 4. First Aid Measures

Eye Contact	Hold eye open and rinse slowly and gently with water for 15-20 minutes. Remove contact lenses, if present, after the first 5 minutes, then continue rinsing eyes. Get medical attention immediately.
Skin Contact	Take off contaminated clothing. Flush immediately with plenty of water for at least 15 minutes. Get medical attention immediately.
Inhalation	If breathing is affected, remove to fresh air. If person is not breathing, call 911 or an ambulance and then give artificial respiration, preferably by mouth to mouth, if possible. Get medical attention immediately.
Ingestion	If ingested, call a physician or Poison Control Center immediately. Have person sip a glass of water if able to swallow. Do not induce vomiting unless told to do so by a poison control center or doctor. Never give anything by mouth to an unconscious person.
Notes to Physician	Probable mucosal damage may contraindicate the use of gastric lavage.

Section 5. Fire Fighting Measures

Flammability of the Product	None known.
Flash Points	Closed cup: >93.333°C (200°F).
Products of Combustion	None known.
Fire Fighting Media and Instructions	Extinguish with water spray or carbon dioxide, dry chemical powder or appropriate foam. Normal fire fighting procedure may be used.
Protective Clothing (Fire)	Put on appropriate personal protective equipment (see Section 8).
Special Remarks on Fire and Explosion Hazards	Corrosive material (See sections 8 and 10).

Section 6. Accidental Release Measures

Personal Precautions	Put on appropriate personal protective equipment (see Section 8).
Environmental Precautions and Clean-up Methods	In the event of major spillage: Use appropriate containment to avoid environmental contamination. Sweep or scrape up material. Place in suitable clean, dry containers for disposal by approved methods. Use a water rinse for final clean-up

Section 7. Handling and Storage

Handling	Avoid contact with eyes, skin and clothing. Remove and wash contaminated clothing and footwear before re-use. Wash thoroughly after handling. Avoid breathing vapors or spray mists. Product residue may remain on/in empty containers. All precautions for handling the product must be used in handling the empty container and residue. Do not taste or swallow. FOR COMMERCIAL AND INDUSTRIAL USE ONLY.
Storage	Store in a dry, cool and well-ventilated area. Protect from freezing. Keep container tightly closed. KEEP OUT OF REACH OF CHILDREN.

Continued on Next Page

Page: 2/4

Safety Data Sheet

J-512 SANITIZER

Section 8. Exposure Controls/Personal Protection

Engineering Controls	Good general ventilation should be sufficient to control airborne levels. Respiratory protection is not required if good ventilation is maintained.
Personal Protection	
Eyes	Chemical splash goggles.
Hands	Chemical resistant gloves.
Respiratory	No specific personal protection equipment is required.
Feet	No specific personal protection equipment is required.
Body	If major exposure is possible, wear suitable protective clothing and footwear.

Section 9. Physical and Chemical Properties

Physical State and Appearance	Liquid.
Odor	Bland.
Color	Clear Pink.
pH	6 to 8 [Neutral.]
Specific Gravity	1.02
Boiling/Condensation Point	>93°C (199.4°F)
Melting/Freezing Point	<0°C (32°F)
Solubility in water	Complete.

Section 10. Stability and Reactivity

Stability and Reactivity	The product is stable.
Conditions of Instability	None known.
Incompatibility with Various Substances	None known.
Hazardous Decomposition Products	None known.
Hazardous Polymerization	Will not occur.

Section 11. Toxicological Information

Acute toxicity	Corrosive. Acute oral toxicity (LD50): Estimated to be between 500 and 5000 mg/kg (rat).
Effects of Chronic Exposure	None known.
Other Toxic Effects	None known.

Section 12. Ecological Information

Not available.

Section 13. Disposal Considerations

Waste Information	Handle as a Pesticide waste. Do not bury. Do not dispose of with Commercial or Household waste. Dispose of according to all federal, state and local regulations.

Continued on Next Page

Safety Data Sheet

J-512 SANITIZER

Section 14. Transport Information

DOT Classification	
DOT Proper Shipping Name	Please refer to the Bill of Lading/receiving documents for up to date shipping information.
TDG Classification	
TDG Proper Shipping Name	Please refer to the Bill of Lading/receiving documents for up to date shipping information.
TDG Class	

Section 15. Regulatory Information

Reporting in this section is based on ingredients disclosed in Section 2

US Regulations	
	This product is not subject to the reporting requirements under California's Proposition 65.
Registered Product Information	EPA Registration Number: 1839-86-70627
Canadian Regulations	
WHMIS Classification	Exempt
WHMIS Icon	
Registered Product Information	Not applicable.
Chemical Inventory Status	All ingredients of this product are listed or are excluded from listing on the U.S. Toxic Substances Control Act (TSCA) Chemical Substance Inventory

Section 16. Other Information

Other Special Considerations	SDS Serial Range: 003-004.
Version	2

Notice to Reader

This document has been prepared using data from sources considered technically reliable. It does not constitute a warranty, express or implied, as to the accuracy of the information contained within. Actual conditions of use and handling are beyond seller's control. User is responsible to evaluate all available information when using product for any particular use and to comply with all Federal, State, Provincial and Local laws and regulations.

Page: 4/4

1. What is the product name?

2. What is the product use?

3. What is the manufacturer name?

4. List the manufacturer address.

5. What is the manufacturer phone number?

6. What is the validation date?

7. List three handling precautions indicated on the SDS.

8. List two storage precautions indicated on the SDS.

9. Identify three hazardous routes of entry.

10. List any potential acute health effects to eyes.

11. List any potential acute health effects to skin.

12. List any potential acute health effects from inhalation.

13. List any potential acute health effects from ingestion.

14. Are any fire hazards identified?

15. Are any reactivity hazards identified?

16. Describe first-aid measures for eye contact.

17. Describe first-aid measures for skin contact.

18. Describe first-aid measures for inhalation.

19. What components/ingredients are identified?

20. Are any exposure limits to components identified?

21. List four pieces of personal protective equipment required for handling this product.

SECTION 3.1 KNIFE CONSTRUCTION

Name ______________________ **Date** ______________

True-False

T F **1.** A rat-tail tang runs the length of the knife handle but is not as wide as the handle.

T F **2.** Ceramic blades provide a sharper edge for a longer period than any other material.

T F **3.** A knife with a dull edge is safer than a knife with a sharp edge.

T F **4.** Knife blades constructed from stainless steel do not discolor or react with acidic foods.

T F **5.** Serrated edge blades are the most common type of knife blade.

Multiple Choice

______________ **1.** ___ knife handles are becoming less common due to their lack of durability and how easily they trap bacteria.
- A. Stainless steel
- B. Wooden
- C. Ceramic
- D. Plastic

______________ **2.** The tip is the front ___ of the knife blade.
- A. eighth
- B. quarter
- C. third
- D. half

______________ **3.** The ___ is the unsharpened tail of a knife blade that extends into the handle.
- A. rivet
- B. spine
- C. tang
- D. edge

______________ **4.** Most knives currently used in the professional kitchen are made of ___ stainless steel or ceramic material.

A. low-density
B. high-density
C. low-carbon
D. high-carbon

______________ **5.** ___ knives discolor over time if they come in contact with highly acidic foods such as tomatoes or lemons.

A. Stainless steel
B. Carbon steel
C. Ceramic
D. High-carbon stainless steel

______________ **6.** The purpose of the bolster is to provide strength to the ___ and prevent food from entering the seam between the blade and the handle.

A. blade
B. handle
C. tang
D. rivets

Completion

______________ **1.** The ___ is the unsharpened top part of a knife blade that is opposite the edge.

______________ **2.** The ___ is the rear portion of a knife blade and is most often used to cut thick items where more force is required.

______________ **3.** Granton edge blades have hollowed out grooves running along both sides that reduce the amount of ___ as the edge of the blade cuts the food.

______________ **4.** ___ blades are thinner, lighter blades cut from a flat sheet of metal and then ground to form a sharp edge.

______________ **5.** A(n) ___ is a metal fastener used to attach the tang of a knife to the handle.

Name ______________________________ Date ____________________

True-False

T F **1.** A mandoline is a manual cutting tool with adjustable steel blades used to cut food into consistently thin slices.

T F **2.** A clam knife is a special knife with a short, dull-edged blade with a tapered point that is used to open clams.

T F **3.** Slicers are available with a straight, serrated, or granton blade edge.

T F **4.** A paring knife is a short knife with a stiff 2–4 inch blade used to trim fat off meat.

T F **5.** A santoku knife often has a granton edge blade, which prevents food from sticking to the blade.

Multiple Choice

______________ **1.** A ___ is a cross between a chef's knife and a paring knife.
- A. scimitar
- B. cleaver
- C. santoku knife
- D. utility knife

______________ **2.** A ___ knife is a thin knife with a pointed 5–6 inch blade that is either stiff or flexible.
- A. bread
- B. utility
- C. paring
- D. boning

______________ **3.** A butcher's knife is used to cut, section, and portion ___ meats.
- A. raw
- B. cooked
- C. frozen
- D. boiled

______ **4.** A(n) ___ knife is a short knife with a curved blade that is primarily used to carve vegetables into a seven-sided football shape with flat ends.

A. paring
B. tourné
C. clam
D. oyster

______ **5.** A ___ is a nearly flat, razor-sharp, handheld grater that shaves food into fine or very fine pieces.

A. channel knife
B. peeler
C. rasp grater
D. mandoline

Completion

______ **1.** A bread knife is a knife with a(n) ___ blade 8–12 inches long.

______ **2.** A cleaver is a heavy, rectangular-bladed knife that is used to cut through thick meat and ___.

______ **3.** A(n) ___ is a special cutting tool that has a half-ball cup with a blade edge attached to a handle and is used to cut fruits and vegetables into uniform spheres.

______ **4.** To use a(n) ___, the cutting holes are drawn across the peel of a citrus fruit.

______ **5.** A(n) ___ knife, also known as a French knife, is a versatile knife used for slicing, dicing, and mincing.

______ **6.** A(n) ___ is a special cutting tool with a swiveling, double-edged blade that is attached to a handle and is used to remove the skin or peel from fruits and vegetables.

Name ______________________________ **Date** ____________________

True-False

T F **1.** With the proper knife grip and hand position, a rocking motion is used to cut with a chef's knife.

T F **2.** A whetstone is used to sharpen professional knives.

T F **3.** Knives should be honed after each use to maintain smooth, sharp edges.

T F **4.** When properly holding a knife, the side of the blade should rest against the knuckle of the middle finger of the guiding hand.

T F **5.** When walking with a knife, the knife should always be kept pointing down and it should be held across the front of the body.

T F **6.** Knives should be sharpened at a 50° angle.

T F **7.** Knives should be stored in sleeves, guards, or knife holders to avoid injury.

T F **8.** Injury is more likely to occur with a sharp knife than a dull one.

Multiple Choice

______________ **1.** Always pass a knife to a person ___.

A. blade first
B. handle first
C. by sliding it across a table
D. by placing it in a holder

______________ **2.** When sharpening a knife, the knife blade is slowly dragged across the stone from ___ while applying light pressure.

A. heel to tip
B. tip to heel
C. edge to spine
D. spine to edge

______________ **3.** To correctly position the fingers of the guiding hand, imitate the shape of a(n) ___ on the table.
- A. spider
- B. crustacean
- C. disk
- D. egg

______________ **4.** A ___ is used to grind the edge of a blade to the proper angle for sharpness.
- A. steel
- B. whetstone
- C. bolster
- D. tang

______________ **5.** Honing is the process of aligning a blade's ___ and removing any burrs or rough spots on the blade.
- A. handle
- B. spine
- C. edge
- D. point

Completion

______________ **1.** Food items are always cut on a nonporous cutting board because the nonporous surface greatly reduces the risk of ___.

______________ **2.** Knives should never be washed in a commercial dish machine as the heat and chemicals can ruin the ___.

______________ **3.** When using a knife, the more ___ that is applied the higher the risk of the knife slipping and of personal injury occurring.

______________ **4.** The hand not holding the knife is referred to as the ___ hand.

______________ **5.** A(n) ___ should always be used to hone the blade of a knife after sharpening.

Matching

______________ **1.** Whetstone

______________ **2.** Steel

______________ **3.** Honing

A. A steel rod approximately 18 inches long that is attached to a handle and used to align the edge of knife blades

B. The process of aligning a blade's edge and removing any burrs or rough spots on the blade

C. A stone used to grind the edge of a blade to the proper angle for sharpness

Chapter 3 Review
Knife Skills

Section 3.4 Knife Cuts

Name ______________________________ **Date** ____________________

True-False

T F **1.** Shallots, garlic, and fresh herbs are commonly julienned.

T F **2.** Uniform cuts ensure that items cook evenly.

T F **3.** To make a diagonal cut, an item is placed at a 45° angle to the knife blade.

T F **4.** When dicing an onion, the root end is not cut off completely.

T F **5.** A fine julienne cut produces a stick-shaped item ⅛ × ⅛ × 2 inches long.

T F **6.** A fine brunoise is cut from a batonnet cut.

T F **7.** A chiffonade cut produces thin shreds of fruits and root vegetables.

T F **8.** In slicing, the knife is pulled backward or slid forward through the item.

T F **9.** Chopping is rough-cutting an item so that there are relatively small pieces throughout and complete uniformity in shape and size.

T F **10.** A tourné cut produces a seven-sided, football-shaped item with small, flat ends.

Multiple Choice

______________ **1.** In comparison with the diagonal cut, the oblique cut produces larger pieces that are more ___.

A. wedge-shaped
B. square-shaped
C. circular
D. oblong

______________ **2.** A(n) ___ cut is a dice cut that produces a cube-shaped item with six equal sides measuring ⅛ inch each.

A. oblique
B. julienne
C. paysanne
D. brunoise

_______________ **3.** To make a(n) ___ cut, the cylindrical food is placed perpendicular to the knife blade and then sliced to create disks.

A. chiffonade
B. rondelle
C. batonnet
D. oblique

_______________ **4.** All ___ cuts begin by squaring off the item to be cut.

A. diagonal
B. chiffonade
C. stick
D. fluted

_______________ **5.** A ___ cut is a dice cut that produces a cube-shaped item with six equal sides measuring ¼ inch each.

A. brunoise
B. paysanne
C. small dice
D. julienne

_______________ **6.** A(n) ___ cut is a slicing cut that produces long strips of leafy greens or herbs.

A. fine julienne
B. oblique
C. chiffonade
D. brunoise

_______________ **7.** ___ cuts are made from cylindrical food items that are cut on the bias.

A. Rondelle
B. Diagonal
C. Julienne
D. Tourné

_______________ **8.** A fine brunoise cut is a dice cut that produces a cube-shaped item with six equal sides measuring ___ inch each.

A. 1⁄16
B. ⅛
C. ¼
D. ½

_______________ **9.** ___ are often fluted.

A. Carrots
B. Onions
C. Artichokes
D. Button mushrooms

_______________ **10.** When making an oblique cut, the item is rolled ___° after each slice to produce two angled sides.
A. 45
B. 90
C. 180
D. 360

Completion

_______________ **1.** A(n) ___ cut is a stick cut that produces a stick-shaped item ¼ × ¼ × 2 inches long.

_______________ **2.** Dice cuts are precise cubes cut from uniform ___ cuts.

_______________ **3.** A(n) ___ cut is a dice cut that produces a flat-square, round, or triangular cut that is ½ × ½ × ⅛ inch thick.

_______________ **4.** A(n) ___ cut is a specialty cut that produces a decorative spiral pattern on the surface of an item by removing only a sliver with each cut.

_______________ **5.** A(n) ___ dice is ½ × ½ × ½ inch cubes.

_______________ **6.** Mincing and ___ have fewer applications in the professional kitchen than the other knife cuts.

_______________ **7.** A rondelle cut, also known as a round cut, is a slicing cut that produces ___.

_______________ **8.** A fine julienne cut is a stick cut that produces a stick-shaped item ___ inches long.

_______________ **9.** Cutting a flat-square, tile-shaped paysanne is done in the same manner as a medium dice, except that each batonnet stick is cut at ___-inch intervals.

_______________ **10.** ___ is the process of finely chopping an item to yield very small pieces that are not entirely uniform in shape.

Matching

_______________ **1.** Julienne cut

_______________ **2.** Mincing

_______________ **3.** Rondelle cut

_______________ **4.** Dice cut

_______________ **5.** Tourné cut

_______________ **6.** Paysanne cut

_______________ **7.** Chiffonade cut

_______________ **8.** Batonnet cut

_______________ **9.** Diagonal cut

_______________ **10.** Fluted cut

_______________ **11.** Brunoise cut

_______________ **12.** Oblique cut

_______________ **13.** Fine julienne

_______________ **14.** Fine brunoise

_______________ **15.** Chopping

A. A specialty cut that produces a decorative spiral pattern on the surface of an item by removing only a sliver of the item with each cut

B. A slicing cut that produces flat-sided, oval slices

C. A stick cut that produces a stick-shaped item ¼ × ¼ × 2 inches long

D. A cut that produces a flat-square, round, or triangular cut ½ × ½ × ⅛ inch thick

E. The process of finely chopping an item to yield very small pieces that are not entirely uniform in shape

F. A stick cut that produces a stick-shaped item ⅛ × ⅛ × 2 inches long

G. A slicing cut that produces wedge-shaped pieces with two angled sides

H. A dice cut that produces a cube-shaped item with six equal sides measuring ⅛ inch each

I. A specialty cut that produces a seven-sided, football-shaped item with small, flat ends

J. A slicing cut that produces disks

K. A slicing cut that produces long strips of leafy greens or herbs

L. A stick cut that produces a stick-shaped item 1⁄16 × 1⁄16 × 2 inches long.

M. The process of rough-cutting an item into small pieces that lack uniformity in shape and size

N. A dice cut that produces a cube-shaped item with six equal sides measuring 1⁄16 inch each

O. Precise cubes cut from uniform stick cuts

Name ______________________ **Date** ______________

Activity: Identifying Knives and Special Cutting Tools

A chef uses many different types of knives and special cutting tools in the professional kitchen. Knowing which knife or special cutting tool to use in a given application makes working with knives safer and more efficient.

Identify each large knife based on appearance.

______ **1.** Slicer

______ **2.** Butcher's knife

______ **3.** Chef's knife

______ **4.** Bread knife

______ **5.** Scimitar

______ **6.** Utility knife

______ **7.** Flexible (straight) boning knife

______ **8.** Cleaver

______ **9.** Santoku knife

______ **10.** Stiff (curved) boning knife

Mercer Cutlery
Ⓐ

Mercer Cutlery
Ⓑ

Canada Cutlery Inc.
Ⓒ

Mercer Cutlery
Ⓓ

Mercer Cutlery
Ⓔ

Mercer Cutlery
Ⓕ

Mercer Cutlery
Ⓖ

Mercer Cutlery
Ⓗ

Mercer Cutlery
Ⓘ

Messermeister
Ⓙ

Identify each small knife based on appearance.

______________ **11.** Oyster knife

______________ **12.** Paring knife

______________ **13.** Clam knife

______________ **14.** Tourné knife

Mercer Cutlery

(K)

Browne-Halco (NJ)

(N)

American Metalcraft, Inc.

(L)

Mercer Cutlery

(M)

Identify each special cutting tool based on appearance.

______________ **15.** Rasp Grater

______________ **16.** Zester

______________ **17.** Channel knife

______________ **18.** Peeler

______________ **19.** Mandoline

______________ **20.** Parisienne scoop

Dexter-Russell, Inc.

(O)

Carlisle FoodService Products

(P)

Paderno World Cuisine

(Q)

Dexter-Russell, Inc.

(R)

Messermeister

(S)

Paderno World Cuisine

(T)

Activity: Analyzing Knife Safety

Because of their sharp edges, knives are dangerous and improper use can lead to injury. Safety precautions such as holding, using, carrying, washing, and storing knives properly should be followed.

Complete the following:

1. Describe how to hold a chef's knife using a pinch grip.

2. Describe how to safely position the guiding hand into a claw grip.

3. Explain why food items should always be cut on a nonporous cutting board.

4. Describe how to safely store knives.

5. Explain why applying a large amount of pressure when using a knife is unsafe.

6. Explain why knives should never be washed in a commercial dish machine.

7. Describe how to safely pass a knife to another person.

Activity: Sharpening and Honing Knives

A sharp knife is an essential tool in culinary arts. Having a sharp knife helps prevent injury because less pressure is required to use a sharp knife compared to a dull knife. Sharpening and honing knives is an important part of regular mise en place.

Complete the following procedures for sharpening knives.

______________ **1.** Use the ___-grit side of a whetstone and place the heel of the knife near the end of the stone, tilting the spine to form the desired angle.

______________ **2.** Draw the knife blade across the length of the whetstone in a continuous arclike motion, starting with the knife ___ and ending with the knife ___. Repeat this step 8 to 10 times.

______________ **3.** Repeat steps 1 and 2 on the ___ side of the knife blade.

______________ **4.** Turn the whetstone over to the ___-grit side and repeat the sharpening procedure.

______________ **5.** With the blade facing ___ the body, use a clean, folded towel to wipe residue from the knife blade.

Complete the following procedures for honing knives.

______________ **6.** Place the heel of a knife at the desired ___ along one side of the steel near the handle.

______________ **7.** Slide the knife down the steel, moving the blade in an arc along the steel. Finish the stroke with the ___ of the knife at the bottom of the steel.

______________ **8.** Place the heel of the knife at the desired angle along the other side of the ___ near the handle and repeat the previous step.

______________ **9.** Repeat the honing procedure 3 to 5 times, using the ___ number of strokes on each side of the blade.

______________ **10.** With the blade facing away from the body, use a(n) ___ to wipe residue from the knife blade.

Sharpen and hone a knife using proper procedures. Have an instructor check the result and sign on the line below when complete.

11. Sharpening Proficiency

Instructor:______________________________ Date:______________

12. Honing Proficiency

Instructor:______________________________ Date:______________

Activity: Purchasing Knives

Care and consideration should be taken when purchasing knives. The chef's hand size, amount of expected use, and cost are all factors in choosing knives.

Complete the following sentences and answer the questions that follow for each knife listed. Any local or online vendor may be used to acquire the information required. Provide sources for pricing.

Chef's Knife

_______________ **1.** The manufacturer is ___.

_______________ **2.** The handle is constructed from ___.

_______________ **3.** The blade is constructed from ___.

_______________ **4.** Is the blade forged or stamped?

_______________ **5.** The length of the blade is ___.

_______________ **6.** The purchase price is $___.

Utility Knife

_______________ **7.** The manufacturer is ___.

_______________ **8.** The handle is constructed from ___.

_______________ **9.** The blade is constructed from ___.

_______________ **10.** Is the blade forged or stamped?

_______________ **11.** The length of the blade is ___.

_______________ **12.** The purchase price is $___.

Paring Knife

_______________ **13.** The manufacturer is ___.

_______________ **14.** The handle is constructed from ___.

_______________ **15.** The blade is constructed from ___.

_______________ **16.** Is the blade forged or stamped?

_______________ **17.** The length of the blade is ___.

_______________ **18.** The purchase price is $___.

Santoku Knife

__________ **19.** The manufacturer is ___.

__________ **20.** The handle is constructed from ___.

__________ **21.** The blade is constructed from ___.

__________ **22.** Is the blade forged or stamped?

__________ **23.** The length of the blade is ___.

__________ **24.** The purchase price is $___.

Bread Knife

__________ **25.** The manufacturer is ___.

__________ **26.** The handle is constructed from ___.

__________ **27.** The blade is constructed from ___.

__________ **28.** Is the blade forged or stamped?

__________ **29.** The length of the blade is ___.

__________ **30.** The purchase price is $___.

Activity: Reporting on Purchasing Knives

Researching and comparing information is essential when considering what knives to purchase. The data gathered through research helps ensure individual requirements of the foodservice operation are met and the appropriate knives are purchased.

Complete the following:

1. Prepare and present a report about purchasing knives using information gathered from Activity: Purchasing Knives. Use visuals, such as pictures, spreadsheets, manufacturer brochures, etc., to enhance the report.

Chapter 4 Review
Tools and Equipment

Section 4.1 Hand Tools

Name ______________________ **Date** ______________

True-False

T F **1.** A rotary grater is a stainless steel box with grids of various sizes on each side that are used to cut food into small pieces.

T F **2.** A china cap is a perforated, cone-shaped metal strainer that is used to strain gravies, soups, stocks, sauces, and other liquids.

T F **3.** A balloon whisk has very flexible wires, allowing the user to whip a great amount of air into items such as egg whites.

T F **4.** Metal rings and molds are often used to shape foods, such as eggs, as they cook.

T F **5.** A dough docker is a round tool that has wire guides that leave marks indicating where to cut pies, round cakes, or pizzas into equal portions.

T F **6.** A mixing paddle is a long-handled paddle used to stir foods in deep pots or steam kettles.

T F **7.** Volume measures often consist of ¼ cup, ⅓ cup, ½ cup, ¾ cup, and 1 cup measures.

Multiple Choice

______________ **1.** A ___ is a hand-cranked sieve with a bowl-shaped body that is used to purée soft or cooked foods.
A. food mill
B. ricer
C. drum sieve
D. sifter

______________ **2.** ___ allow fat to drain off foods before serving and are thin and flexible.
A. Graters
B. Peels
C. Solid offset spatulas
D. Slotted offset spatulas

______________ **3.** A ___ is a tapered bowl attached to a short tube that is used to transfer substances from one container to another container without spilling.
A. scoop
B. ladle
C. funnel
D. spoodle

______________ **4.** A ___ is a manual slicing tool with adjustable steel blades used to cut food in consistently thin slices.
A. grinder
B. mandoline
C. rasp grater
D. rotary grater

______________ **5.** A ___ is a skimmer with an open-wire design that makes it perfect for removing hot foods from a fryer.
A. colander
B. chinois
C. tamis
D. spider

Completion

______________ **1.** A(n) ___ is a stainless steel scoop of a specific size attached to a handle with a thumb-operated release lever.

______________ **2.** A(n) ___ is a dough-cutting tool with a rotating disk attached to a handle.

______________ **3.** A(n) ___ is a scraping tool consisting of a rubber or silicone blade attached to a long handle that is used to mix foods and to scrape food from bowls, pots, and pans.

______________ **4.** A(n) ___ is a stainless steel spoon used to measure a small volume of an ingredient.

______________ **5.** The number on a portion-controlled scoop indicates the number of level scoopfuls that equal ___ qt.

Matching

______________ **1.** Hand tool

______________ **2.** Kitchen shears

______________ **3.** Palette knife

______________ **4.** Strainer

______________ **5.** Bowl scraper

______________ **6.** Funnel

A. Heavy-gauge scissors used to trim foods during the preparation process

B. A flat, narrow knife with a rounded, 3½–12 inch blade that varies in flexibility

C. Any of a variety of manual tools used to cut, shape, measure, strain, sift, mix, blend, turn, or lift food items

D. A bowl-shaped woven mesh screen, often with a handle, that is used to strain or drain foods

E. A tapered bowl that is attached to a short tube and used to transfer substances from one container to another without spilling

F. A curved, flexible scraping tool that is used to scrape batter or dough out of curved containers

Chapter 4 Review
Tools and Equipment

Section 4.2 Cookware

Name ______________________________ **Date** ____________________

True-False

T F **1.** A bain-marie is a rectangular pan with 4–5 inch sides.

T F **2.** Saucepans are used to cook small amounts of food in a liquid.

T F **3.** Hotel pans can be used to cook foods in an oven or on an open burner.

T F **4.** A double boiler is commonly used to melt chocolate or to make a hollandaise sauce.

Multiple Choice

______________ **1.** A ___ is a small skillet that has very short, sloped sides and is used to prepare crêpes.
A. saucepan
B. cast-iron skillet
C. crêpe pan
D. wok

______________ **2.** A ___ is a large, round, high-walled pot that is taller than it is wide.
A. stockpot
B. rondeau
C. wok
D. poacher

______________ **3.** A ___ is a round, stainless steel food-storage container with high walls used for holding sauces or soups in a water bath or steam table.
A. rondeau
B. bain-marie insert
C. steamer
D. double boiler

______________ **4.** A standard ___ pan is approximately 9 × 5 × 2½ inches.
A. loaf
B. springform
C. pie
D. tart

______________ **5.** A ___ pan is a round pan with a metal clamp on the side that allows the bottom of the pan to be separated from the sides.

A. hotel
B. muffin
C. sheet
D. springform

Completion

______________ **1.** A(n) ___ is a round-bottom pan that is used to stir-fry, steam, braise, stew, or deep fry foods.

______________ **2.** A(n) ___ is a wide, shallow-walled round pot that is used for braising, stewing, and searing meats.

______________ **3.** A(n) ___ is a sauté pan with straight sides.

______________ **4.** A(n) ___ is a small stockpot.

______________ **5.** A(n) ___ is a round or rectangular, shallow baking pan with sloped sides that are smooth or fluted and may have a removable bottom.

Name ______________________________ Date ____________________

True-False

T F 1. Extension cords should never be used to operate commercial equipment.

T F 2. Do not read manufacturer instructions for safe equipment operation prior to use.

T F 3. The NSF logo indicates that the product has a smooth, nonporous, nontoxic, corrosion-resistant surface.

T F 4. Do not unplug equipment before cleaning or disassembling.

T F 5. Unsafe or careless operation of equipment can lead to serious injury.

Multiple Choice

__________ 1. ___ is an organization focused on standards development, product certification, education, and risk management for public health and safety.
A. NSF International
B. ASME
C. EPA
D. OSHA

__________ 2. After cleaning equipment, items should be ___ to prevent foodborne illness.
A. dried with a towel
B. stored
C. sanitized
D. air-dried

__________ 3. If equipment malfunctions or damage is detected, immediately ___.
A. disassemble the equipment
B. notify a supervisor
C. clean the equipment
D. install safety features

__________ 4. The ___ area of the professional kitchen is where used items are cleaned and sanitized.
A. receiving
B. preparation
C. storage
D. sanitation

Completion

________________ **1.** Commercial-grade equipment is designed according to ___ sanitation standards.

________________ **2.** The equipment used in a professional kitchen must be ___ grade to withstand wear and tear and must be able to be easily cleaned.

________________ **3.** The professional kitchen is typically divided into several major areas including safety, receiving, storage, sanitation, and ___.

________________ **4.** The presence of bacteria on equipment in the professional kitchen can create cross-___ and spread foodborne illnesses.

________________ **5.** The ___ mark indicates that a product meets standards such as rounded and smooth external edges.

Chapter 4 Review
Tools and Equipment

Section 4.4 Safety Areas

Name ______________________________ **Date** ____________________

True-False

T F **1.** When removing a candy/deep-fat thermometer from a hot substance, care must be taken to avoid placing it into something cold, as the sudden temperature change can damage the thermometer.

T F **2.** Class C fire extinguishers are used specifically for grease fires.

T F **3.** Standards for safety areas are established by national health departments to help foodservice operations serve food safely and to provide a safe workplace for employees.

T F **4.** Grease must be cleaned off of the ventilation hood and the fire-suppression components regularly to prevent potential fires.

Multiple Choice

______________ **1.** The ___ area is the area where ventilation systems, fire extinguishers, and first aid kits are located.

A. preparation
B. safety
C. sanitation
D. storage

______________ **2.** A(n) ___ thermometer has a clip that allows it to be clipped to the side of a pot during cooking.

A. HACCP
B. instant-read
C. candy/deep-fat
D. electronic probe

______________ **3.** A(n) ___ is a measuring tool that indicates the amount of time that has passed or sounds an alarm when a specified time period has ended.

A. timer
B. thermometer
C. probe
D. analog gauge

______________ **4.** Class ___ fire extinguishers are used to extinguish electrical fires.
A. A
B. B
C. C
D. K

Completion

______________ **1.** An infrared thermometer is a thermometer that measures the ___ temperature of an item.

______________ **2.** A(n) ___ system is required over any open-flame cooking surface or combustible surface such as a gas cooktop or fryer.

______________ **3.** The ___ of an instant-read thermometer is briefly inserted into foods during cooking to determine the internal temperature.

______________ **4.** Class ___ fire extinguishers are for fires involving common combustible materials such as trash, wood, or paper.

Matching

______________ **1.** Ventilation system

______________ **2.** Fire-suppression system

______________ **3.** Electronic probe thermometer

______________ **4.** HACCP thermometer

______________ **5.** Fire extinguisher

A. A large exhaust system that draws heat, smoke, and fumes out of the kitchen and into the outside air

B. An automatic fire-extinguishing system that is activated by the intense heat generated by a fire

C. A metal canister that is filled with pressurized dry chemicals, foam, or water

D. A thermometer with a thin, stainless steel stem attached to a handheld device that records and monitors the temperature of critical items and provides corrective actions when necessary

E. A thermocouple thermometer with a thin, stainless steel stem that is attached by wires to a battery-operated readout device

Chapter 4 Review
Tools and Equipment

Section 4.5 Receiving Areas

Name ______________________________ **Date** ______________

True-False

T F **1.** Platform and bench scales are used to weigh large or heavy boxes and bags.

T F **2.** A receiving area should contain shelving units, storage bins, and chillers.

T F **3.** Dating each received item is essential to ensure older items are used before newly delivered items.

Multiple Choice

______________ **1.** The most common thermometer used in the receiving area is a(n) ___ thermometer.
A. infrared
B. electronic probe
C. HACCP
D. instant-read

______________ **2.** Fresh perishable foods should be delivered at a temperature of ___°F or below.
A. 0
B. 32
C. 41
D. 50

______________ **3.** ___, or the heaviness of a substance, is measured using scales.
A. Weight
B. Volume
C. Grade
D. Density

Completion

______________ **1.** New items should always be placed at the back of the storage area, and older items are rotated to the front for accessibility using the ___ stock rotation method.

______________ **2.** When using any type of scale, the scale is always ___, or set to zero, before weighing any items.

____________________ **3.** Stainless steel inspection ___ in the receiving area allow the clerk to open boxes for inspection without having to bend down.

Matching

____________________ **1.** Dolly

____________________ **2.** Pallet jack

____________________ **3.** Utility cart

A. A piece of equipment that is used to move large amounts of stock from the receiving area to storage areas

B. A piece of equipment that makes transporting food, equipment, and other small items from one area to another easier and faster

C. A piece of equipment used to move heavy items or large boxes from one area to another

Name ______________________________ **Date** ________________

True-False

T F **1.** A holding cabinet is an insulated container made of heavy polyurethane that is designed to hold hotel pans of hot or cold foods during transport.

T F **2.** Roll-in refrigeration units can be built to any size or shape.

T F **3.** A proofing cabinet can hold hot food without drying it out.

T F **4.** A chafing dish is a hotel pan inside of a stand with a water reservoir and a portable heat source underneath.

T F **5.** A lowboy is a reach-in refrigerated unit located beneath a work surface.

Multiple Choice

________________ **1.** A(n) ___ is an open-top table with heated wells that are filled with water.
A. insulated carrier
B. steam table
C. bain-marie
D. chafing dish

________________ **2.** All refrigeration units must be kept at ___°F or below.
A. 31
B. 35
C. 41
D. 45

________________ **3.** A(n) ___ is shelving with rails in which cans of product can be loaded from the top.
A. overhead rack
B. wire shelf
C. dunnage rack
D. can rack

_______________ **4.** Shelving units must be a minimum of ___ inches above the floor to allow access for cleaning underneath the unit.
A. 2
B. 4
C. 6
D. 10

_______________ **5.** A ___ is a specialized cooling unit that rapidly reduces the temperature of foods, rendering them safe for immediate storage.
A. chafing dish
B. lowboy
C. bain-marie
D. blast chiller

Completion

_______________ **1.** A(n) ___ is shelving consisting of reinforced platforms that keep heavy items at least 12 inches above the floor.

_______________ **2.** Separate storage areas are designated within the professional kitchen for ___, cold, and hot items.

_______________ **3.** A(n) ___ is a lockable, wire-cage storage unit on wheels used to hold expensive items such as fine china or restricted items such as alcohol.

_______________ **4.** A(n) ___, also known as a tallboy, is a tall cart on wheels with rails that hold entire sheet pans of food.

_______________ **5.** A(n) ___ cabinet is a holding cabinet that contains both temperature and humidity controls.

Matching

_______________ **1.** Hot storage

_______________ **2.** Cold storage

_______________ **3.** Dry storage

A. An area used to hold dry or packaged items that do not require refrigeration

B. An area used to hold hot foods at safe temperatures while serving or until needed for service

C. An area that includes refrigeration and freezer units to hold perishable items at safe temperatures

Name ______________________________ Date ____________________

True-False

T F **1.** In a compartment sink, the first compartment is for a mixture of sanitizing solution and warm water.

T F **2.** Examples of sanitation areas include custodial closets and warewashing stations.

T F **3.** Only high-temperature dish machines are available.

Multiple Choice

_______________ **1.** A ___ dish machine has a door that is raised up to load racks of scraped dirty dishes and glassware.
- A. single-tank
- B. carousel
- C. conveyor
- D. multitank

_______________ **2.** In a four-compartment sink, the ___ compartment is used to rinse off large debris.
- A. first
- B. second
- C. third
- D. fourth

_______________ **3.** Compartment sinks should not be used for ___.
- A. washing glass
- B. rinsing dishes
- C. sanitizing food preparation items
- D. handwashing

Completion

_______________ **1.** A(n) ___ is a food grinder mounted beneath warewashing sinks to eliminate solid food material.

_______________ **2.** The process of cleaning and sanitizing pots, pans, and hand tools requires very ___ water.

_______________ **3.** Warewashing sinks typically have at least ___ compartments.

Matching

__________________ **1.** Single-tank dish machine

__________________ **2.** Multitank dish machine

__________________ **3.** Food waste disposer

A. A grinder into which food is rinsed prior to being loaded into a commercial dishwasher

B. A dish machine in which the wash cycle begins automatically after a rack is loaded and the door is closed

C. A dish machine in which a conveyor takes the racks through the prewash, wash, rinse, sanitize, and dry cycle

Chapter 4 Review
Tools and Equipment

Section 4.8 Preparation Areas

Name ______________________ **Date** ______________

True-False

T F **1.** An immersion blender can be inserted into a saucepot to purée a soup.

T F **2.** An impinger conveyor oven directs heat only from above a food item as it moves along a conveyor belt.

T F **3.** A work section is an area where members of the kitchen staff are all working toward the same goal at the same time.

T F **4.** Because the heat is so intense, flavor or moisture is lost in the foods cooked in a flashbake oven.

T F **5.** The cold foods section of the professional kitchen is also known as a garde manger section.

Multiple Choice

______________ **1.** The ___ attachment for a mixer is used for mixing and creaming.
A. shredder
B. hook
C. whip
D. paddle

______________ **2.** A(n) ___ is an electromagnetic unit that uses a magnetic coil below a flat surface to heat food rapidly.
A. griddle
B. salamander
C. induction cooktop
D. flat-top range

______________ **3.** ___ tables are used throughout the kitchen because they have very durable surfaces and are easy to clean.
A. Stainless steel
B. Aluminum
C. Polyurethane
D. Cast-iron

_______________ **4.** Typical work stations within a ___ section include dough preparation, dessert preparation, and plated desserts.
A. short order
B. hot foods
C. cold foods
D. baking and pastry

_______________ **5.** A ___ generates steam using an internal boiler, which circulates around the food to cook it rapidly.
A. steam table
B. bain-marie
C. pressure steamer
D. convection steamer

_______________ **6.** A(n) ___ is a manual or electric device used to extract juice from citrus fruits.
A. reamer
B. juice extractor
C. immersion blender
D. blender

_______________ **7.** A ___ is a cooking unit consisting of a large metal grate placed over a heat source.
A. range
B. grill
C. broiler
D. griddle

_______________ **8.** A ___ oven is a drawerlike oven that is commonly stacked one on top of another, providing multiple-temperature baking shelves.
A. microwave
B. deck
C. combi
D. cook-and-hold

_______________ **9.** A ___ is a tall appliance with a slender canister that is used to chop, blend, purée, or liquefy food.
A. mixer
B. juicer
C. blender
D. food processor

_______________ **10.** A ___ section typically contains sinks, work tables, slicers, mixers, and a range.
A. prep
B. beverage
C. cold foods
D. short order

Completion

______________ **1.** The ___ section of the professional kitchen is where foods such as hot sandwiches and breakfast items are prepared.

______________ **2.** ___ choppers and vertical cutter/mixers are used to cut large amounts of foods quickly and efficiently.

______________ **3.** A salamander is a small overhead ___ that is usually attached to an open burner range.

______________ **4.** A(n) ___ is a device that is placed in a water bath to keep a uniform temperature for sous vide cooking.

______________ **5.** A(n) ___ oven is a large oven that rotates 10–80 pans of food as it cooks.

______________ **6.** Typical stations within a(n) ___ section include the warewashing station, dish machine station, and maintenance station.

______________ **7.** A(n) ___ has a regulator providing a wide range of slice thicknesses and a feed grip that firmly holds the top of the food item or serves as a pusher plate for slicing small end pieces.

______________ **8.** ___ used in the professional kitchen include convection, combi, deck, infrared, flashbake, wood-fired, smoker, and cook-and-hold.

______________ **9.** A(n) ___ oven is an oven that uses both infrared radiation and light waves to cook foods quickly and evenly from above and below.

______________ **10.** A(n) ___ oven is a gas or electric oven with an interior fan that circulates dry, hot air throughout the cabinet.

Matching

______________ **1.** Work section

______________ **2.** Preparation area

______________ **3.** Work station

______________ **4.** Cold foods section

______________ **5.** Baking and pastry section

______________ **6.** Prep section

______________ **7.** Hot foods section

______________ **8.** Short order section

A. An area in the professional kitchen where food items are prepared and cooked

B. An area within a work section where specific tasks are performed by specific people

C. An area where members of the kitchen staff are all working toward the same goal at the same time

D. The section of the kitchen where food items that are needed by one or more work stations are prepped

E. Typical work stations within the section include salads, sandwiches, and plated desserts

F. The section of the professional kitchen where foods such as hot sandwiches and breakfast items are prepared

G. The section in which scales, mixers, proofing cabinets, sheeters, ovens, refrigeration units, and special work tables are located

H. The section of the kitchen where foods are actually cooked

Chapter 4 Activities
Tools and Equipment

Name ______________________________ **Date** ____________________

Activity: Identifying Tools and Equipment

Each area of the kitchen contains the appropriate tools and equipment needed to perform assigned tasks. In order to uphold standards and meet the needs of guests, it is important to be able to identify and use tools and equipment efficiently and safely.

Identify the name of the tool or piece of equipment and describe how it is used.

__________________ **1.** The item pictured is a(n) ___.

2. How is the item used in the professional kitchen?

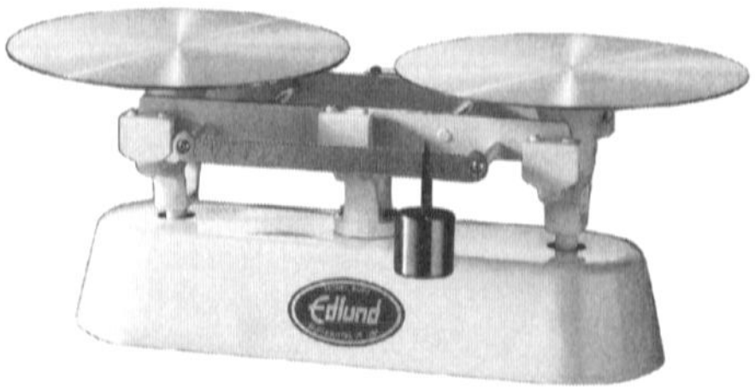

Edlund Co.

__________________ **3.** The item pictured is a(n) ___.

4. How is the item used in the professional kitchen?

Carlisle FoodService Products

__________________ **5.** The item pictured is a(n) ___.

6. How is the item used in the professional kitchen?

Edlund Co.

______________ **7.** The item pictured is a(n) ___.

8. How is the item used in the professional kitchen?

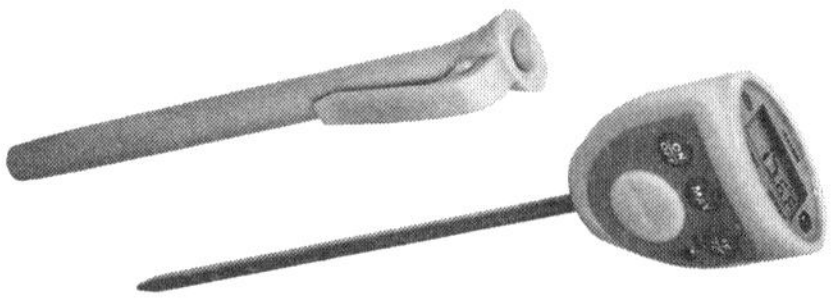

Cooper-Atkins Corporation

______________ **9.** The item pictured is a(n) ___.

10. How is the item used in the professional kitchen?

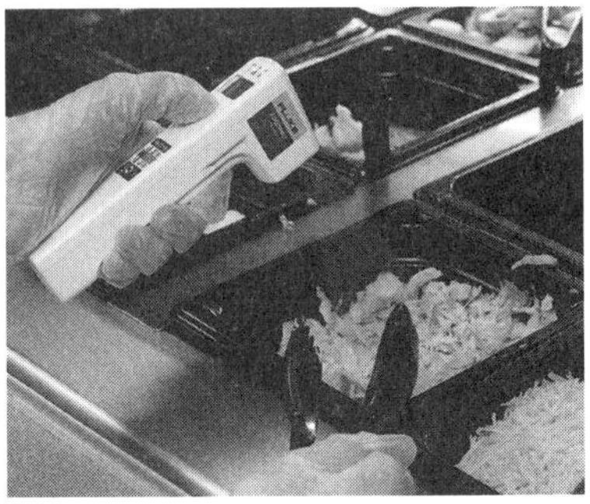

Fluke Corporation

______________ **11.** The items pictured are ___.

12. How are the items used in the professional kitchen?

Carlisle FoodService Products

______________ **13.** The items pictured are ___.

14. How are the items used in the professional kitchen?

Carlisle FoodService Products

______________ **15.** The items pictured are ___.

16. How are the items used in the professional kitchen?

Browne-Halco (NJ)

_______________ **17.** The item pictured is a(n) ___.

18. How is the item used in the professional kitchen?

_______________ **19.** The items pictured are ___.

20. How are the items used in the professional kitchen?

Carlisle FoodService Products

_______________ **21.** The item pictured is a(n) ___.

22. How is the item used in the professional kitchen?

Carlisle FoodService Products

_______________ **23.** The items pictured are ___.

24. How are the items used in the professional kitchen?

Carlisle FoodService Products

_______________ **25.** The items pictured are ___.

26. How are the items used in the professional kitchen?

Carlisle FoodService Products

__________________ **27.** The item pictured is a(n) ___.

28. How is the item used in the professional kitchen?

Matfer Bourgeat USA

__________________ **29.** The item pictured is a(n) ___.

30. How is the item used in the professional kitchen?

American Metalcraft, Inc.

__________________ **31.** The item pictured is a(n) ___.

32. How is the item used in the professional kitchen?

Browne-Halco (NJ)

__________________ **33.** The item pictured is a(n) ___.

34. How is the item used in the professional kitchen?

Browne-Halco (NJ)

__________________ **35.** The item pictured is a(n) ___.

36. How is the item used in the professional kitchen?

All Clad Metalcrafters

____________________ **37.** The items pictured are ___.

38. How are the items used in the professional kitchen?

American Metalcraft, Inc.

____________________ **39.** The items pictured are ___.

40. How are the items used in the professional kitchen?

Carlisle FoodService Products

____________________ **41.** The items pictured are ___.

42. How are the items used in the professional kitchen?

Carlisle FoodService Products

____________________ **43.** The item pictured is a(n) ___.

44. How is the item used in the professional kitchen?

Paderno World Cuisine

____________________ **45.** The item pictured is a(n) ___.

46. How is the item used in the professional kitchen?

Browne-Halco (NJ)

__________________ **47.** The item pictured is a(n) ___.

48. How is the item used in the professional kitchen?

American Metalcraft, Inc.

__________________ **49.** The item pictured is a(n) ___.

50. How is the item used in the professional kitchen?

Dexter-Russell, Inc.

__________________ **51.** The item pictured is a(n) ___.

52. How is the item used in the professional kitchen?

American Metalcraft, Inc.

__________________ **53.** The item pictured is a(n) ___.

54. How is the item used in the professional kitchen?

American Metalcraft, Inc.

__________________ **55.** The item pictured is a(n) ___.

56. How is the item used in the professional kitchen?

Detecto, A Division of Cardinal Scale Manufacturing Co.

____________________ **57.** The item pictured is a(n) ___.

58. How is the item used in the professional kitchen?

PolyScience

____________________ **59.** The item pictured is a(n) ___.

60. How is the item used in the professional kitchen?

Chef's Choice® by EdgeCraft Corporation

____________________ **61.** The item pictured is a(n) ___.

62. How is the item used in the professional kitchen?

Vulcan-Hart, a division of the ITW Food Equipment Group, LLC

____________________ **63.** The item pictured is a(n) ___.

64. How is the item used in the professional kitchen?

Carlisle FoodService Products

____________________ **65.** The item pictured is a(n) ___.

66. How is the item used in the professional kitchen?

InterMetro Industries Corporation

__________________ **67.** The item pictured is a(n) ___.

68. How is the item used in the professional kitchen?

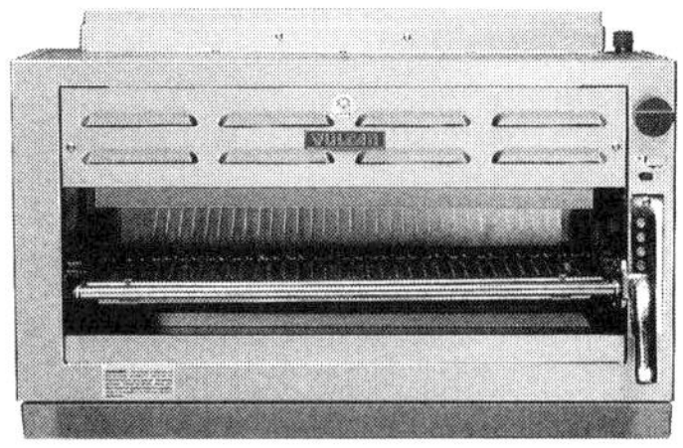

Vulcan-Hart, a division of the ITW Food Equipment Group, LLC

__________________ **69.** The item pictured is a(n) ___.

70. How is the item used in the professional kitchen?

Vita-Mix® Corporation

__________________ **71.** The item pictured is a(n) ___.

72. How is the item used in the professional kitchen?

Cres Cor

__________________ **73.** The item pictured is a(n) ___.

74. How is the item used in the professional kitchen?

Bunn-O-Matic Corporation

__________________ **75.** The item pictured is a(n) ___.

76. How is the item used in the professional kitchen?

American Metalcraft, Inc.

____________________ **77.** The item pictured is a(n) ___.

78. How is the item used in the professional kitchen?

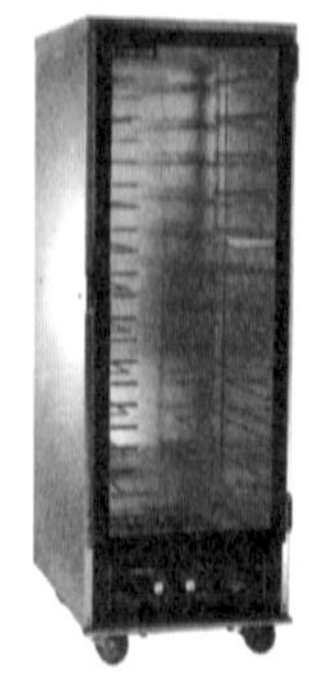

Cres Cor

____________________ **79.** The item pictured is a(n) ___.

80. How is the item used in the professional kitchen?

Hobart

____________________ **81.** The item pictured is a(n) ___.

82. How is the item used in the professional kitchen?

True FoodService Equipment, Inc.

____________________ **83.** The item pictured is a(n) ___.

84. How is the item used in the professional kitchen?

U.S. Range

____________________ **85.** The item pictured is a(n) ___.

86. How is the item used in the professional kitchen?

Vulcan-Hart, a division of the ITW Food Equipment Group LLC

__________________ **87.** The item pictured is a(n) ___.

88. How is the item used in the professional kitchen?

Hobart

__________________ **89.** The item pictured is a(n) ___.

90. How is the item used in the professional kitchen?

Hobart

__________________ **91.** The item pictured is a(n) ___.

92. How is the item used in the professional kitchen?

Vulcan-Hart, a division of the ITW Food Equipment Group LLC

__________________ **93.** The item pictured is a(n) ___.

94. How is the item used in the professional kitchen?

Cres Cor

__________________ **95.** The item pictured is a(n) ___.

96. How is the item used in the professional kitchen?

Detecto, A Division of Cardinal Scale Manufacturing Co.

____________________ **97.** The items pictured are ___.

98. How are the items used in the professional kitchen?

Edlund Co.

____________________ **99.** The item pictured is a(n) ___.

100. How is the item used in the professional kitchen?

Hobart

____________________ **101.** The item pictured is a(n) ___.

102. How is the item used in the professional kitchen?

Vulcan-Hart, a division of the ITW Food Equipment Group, LLC

____________________ **103.** The item pictured is a(n) ___.

104. How is the item used in the professional kitchen?

Hobart

____________________ **105.** The item pictured is a(n) ___.

106. How is the item used in the professional kitchen?

Hobart

____________________ **107.** The item pictured is a(n) ___.

108. How is the item used in the professional kitchen?

Hobart

____________________ **109.** The item pictured is a(n) ___.

110. How is the item used in the professional kitchen?

In-Sink-Erator

Activity: Identifying Portion-Control Measures

Portion control scoops and ladles are essential tools during food preparation and meal service. They help provide consistent portions, reduce waste, and control costs. Each scoop size has an approximate capacity in ounces as well as an equivalent volume in cups or tablespoons. Ladles range in size from ½–32 fl oz.

Answer the following questions that relate to scoop capacity.

Scoop Capacity		
Handle Color	**Scoop No.**	**Volume***
White	6	
Gray	8	
Ivory	10	
Green	12	
Blue	16	
Yellow	20	
Red	24	
Black	30	
Purple	40	

* in fl oz

Carlisle FoodService Products

__________ **1.** How many fluid ounces are contained in a number 30 scoop?

__________ **2.** How many fluid ounces are contained in a scoop with a green handle?

__________ **3.** How many fluid ounces are contained in a scoop with a gray handle?

__________ **4.** How many fluid ounces are contained in a number 20 scoop?

__________ **5.** How many fluid ounces are contained in a scoop with a red handle?

__________ **6.** How many fluid ounces are contained in a number 6 scoop?

__________ **7.** How many fluid ounces are contained in a number 16 scoop?

__________ **8.** How many fluid ounces are contained in a scoop with an ivory handle?

__________ **9.** How many fluid ounces are contained in a scoop with a purple handle?

Answer the following questions related to ladle sizes.

Ladle Sizes	
Ladle Marking	Equivalent Volume
½ fl oz	
1 fl oz	
2 fl oz	
3 fl oz	
4 fl oz	
6 fl oz	
8 fl oz	
12 fl oz	
24 fl oz	
32 fl oz	

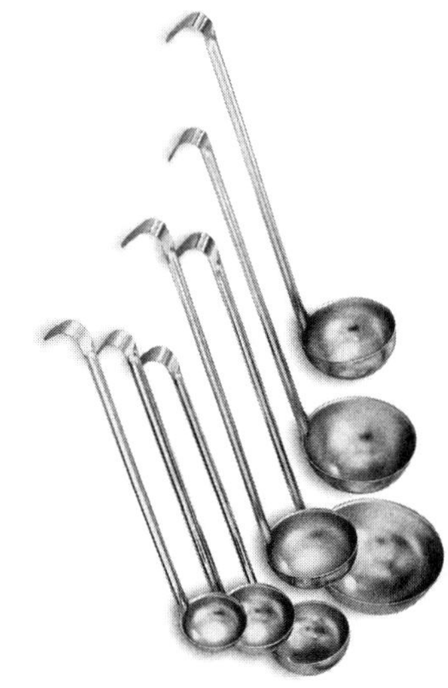

The Vollrath Company, LLC

_______________ **10.** What is the equivalent volume of a ladle with a marking of 6 fl oz?

_______________ **11.** What is the equivalent volume of a ladle with a marking of 2 fl oz?

_______________ **12.** What is the equivalent volume of a ladle with a marking of 24 fl oz?

_______________ **13.** What is the equivalent volume of a ladle with a marking of ½ fl oz?

_______________ **14.** What is the equivalent volume of a ladle with a marking of 8 fl oz?

_______________ **15.** What is the equivalent volume of a ladle with a marking of 4 fl oz?

_______________ **16.** What is the equivalent volume of a ladle with a marking of 32 fl oz?

_______________ **17.** What is the equivalent volume of a ladle with a marking of 3 fl oz?

_______________ **18.** What is the equivalent volume of a ladle with a marking of 12 fl oz?

_______________ **19.** What is the equivalent volume of a ladle with a marking of 1 fl oz?

Activity: Comparing Costs for New and Used Tools and Equipment

Tools and equipment can be a large financial investment. Researching and comparing the cost of new and used equipment can help a foodservice operation find the most appropriate tools and equipment for their specific operation and budget.

Find the new and used price for the following tools and equipment. Any local or online vendor may be used to determine prices. Provide sources for pricing.

____________________ **1.** The purchase price for a new dolly is $___.

____________________ **2.** A used dolly may be purchased for $___.

____________________ **3.** The purchase price for a new inspection table is $___.

____________________ **4.** A used inspection table may be purchased for $___.

____________________ **5.** The purchase price for a new bench scale is $___.

____________________ **6.** A used bench scale may be purchased for $___.

____________________ **7.** The purchase price for a new digital receiving scale is $___.

____________________ **8.** A used digital receiving scale may be purchased for $___.

____________________ **9.** The purchase price for a new wire shelving unit is $___.

____________________ **10.** A used wire shelving unit may be purchased for $___.

____________________ **11.** The purchase price for a new security cage is $___.

____________________ **12.** A used security cage may be purchased for $___.

____________________ **13.** The purchase price for a new speed rack is $___.

____________________ **14.** A used speed rack may be purchased for $___.

____________________ **15.** The purchase price for a new roll-in cold storage unit is $___.

____________________ **16.** A used roll-in cold storage unit may be purchased for $___.

____________________ **17.** The purchase price for a new lowboy is $___.

____________________ **18.** A used lowboy may be purchased for $___.

____________________ **19.** The purchase price for a cold storage reach-in unit is $___.

____________________ **20.** A used cold storage reach-in unit may be purchased for $___.

____________________ **21.** The purchase price for a new steam table is $___.

____________________ **22.** A used steam table may be purchased for $___.

____________________ **23.** The purchase price for a new proofing cabinet is $___.

____________________ **24.** A used proofing cabinet may be purchased for $___.

____________________ **25.** The purchase price for a new automatic coffee urn is $___.

____________________ **26.** A used automatic coffee urn may be purchased for $___.

____________________ **27.** The purchase price for a new stainless steel work table is $___.

____________________ **28.** A used stainless steel work table may be purchased for $___.

____________________ **29.** The purchase price for a new immersion blender is $___.

____________________ **30.** A used immersion blender may be purchased for $___.

____________________ **31.** The purchase price for a new floor mixer is $___.

____________________ **32.** A used floor mixer may be purchased for $___.

____________________ **33.** The purchase price for a new fryer is $___.

____________________ **34.** A used fryer may be purchased for $___.

____________________ **35.** The purchase price for a new open-burner range is $___.

____________________ **36.** A used open-burner range may be purchased for $___.

____________________ **37.** The purchase price for a new griddle is $___.

____________________ **38.** A used griddle may be purchased for $___.

____________________ **39.** The purchase price for a new grill is $___.

____________________ **40.** A used grill may be purchased for $___.

____________________ **41.** The purchase price for a new convection oven is $___.

____________________ **42.** A used convection oven may be purchased for $___.

____________________ **43.** The purchase price for a new deck oven is $___.

____________________ **44.** A used deck oven may be purchased for $___.

____________________ **45.** The purchase price for a new microwave oven is $___.

____________________ **46.** A used microwave oven may be purchased for $___.

____________________ **47.** The purchase price for a new broiler is $___.

____________________ **48.** A used broiler may be purchased for $___.

____________________ **49.** The purchase price for a new salamander is $___.

____________________ **50.** A used salamander may be purchased for $___.

____________________ **51.** The purchase price for a new combi oven is $___.

____________________ **52.** A used combi oven may be purchased for $___.

Activity: Comparing Equipment

There are many factors to consider when purchasing equipment such as costs, warranties, ease of use, and efficiency. Purchasing the right equipment can help promote success. This makes it essential to compare products in order to find items that meet the needs of a particular foodservice operation.

Choose one type of foodservice equipment and contrast three equivalent models. Record the relevant information.

1. What type of foodservice equipment is being compared?

2. Identify the manufacturer of equipment model A.

3. List the name and model number of equipment model A.

4. What is the purchase cost of equipment model A?

5. What, if any, are the delivery charges for equipment model A?

6. What, if any, are the set-up charges for equipment model A?

7. What, if any, is the sales tax applicable to the purchase of equipment model A?

8. Does equipment model A use electricity?

9. Does equipment model A use natural gas or propane? If yes, which type is used?

10. Is plumbing required for equipment model A?

11. What, if any, is the warranty coverage for equipment model A?

12. Is any customer service/product support available for equipment model A?

13. Name any advantages of equipment model A.

14. Name any disadvantages of equipment model A.

15. Identify the manufacturer of equipment model B.

16. List the name and model number of equipment model B.

17. What is the purchase cost of equipment model B?

18. What, if any, are the delivery charges for equipment model B?

19. What, if any, are the set-up charges for equipment model B?

20. What, if any, is the sales tax applicable to the purchase of equipment model B?

21. Does equipment model B use electricity?

22. Does equipment model B use natural gas or propane? If yes, which type is used?

23. Is plumbing required for equipment model B?

24. What, if any, is the warranty coverage for equipment model B?

25. Is any customer service/product support available for equipment model B?

26. Name any advantages of equipment model B.

27. Name any disadvantages of equipment model B.

28. Identify the manufacturer of equipment model C.

29. List the name and model number of equipment model C.

30. What is the purchase cost of equipment model C?

31. What, if any, are the delivery charges for equipment model C?

32. What, if any, are the set-up charges for equipment model C?

33. What, if any, is the sales tax applicable to the purchase of equipment model C?

34. Does equipment model C use electricity?

35. Does equipment model C use natural gas or propane? If yes, which type is used?

36. Is plumbing required for equipment model C?

37. What, if any, is the warranty coverage for equipment model C?

38. Is any customer service/product support available for equipment model C?

39. Name any advantages of equipment model C.

40. Name any disadvantages of equipment model C.

Activity: Reporting on Tools and Equipment Research

The data gathered through research helps ensure individual requirements of the foodservice operation are met and informed decisions are made.

Complete the following:

1. Prepare and present a summary report of purchasing tools and equipment using data gathered from Activity: Comparing Costs for New and Used Tools and Equipment and Activity: Comparing Equipment. Use visuals to enhance the report, such as pictures, spreadsheets, manufacturer brochures, etc.

Activity: Demonstrating Equipment Use

It is imperative to the well-being of employees and guests to know how to use equipment properly. Following manufacturer guidelines for usage, cleaning and sanitizing, and proper storage can help maintain efficient equipment and promote safety for all.

Complete the following:

1. Prepare a demonstration involving the proper use of a tool or a piece of equipment found in the professional kitchen. Explain and demonstrate the purpose and the safe and sanitary operation of the tool or piece of equipment.

Chapter 5 Review
Cost Control Fundamentals

Section 5.1 Standardized Recipes

Name ______________________________ **Date** ________________

True-False

T F **1.** A significant challenge for any foodservice operation is to produce food that is consistent.

T F **2.** The name of a recipe should not reflect the name used on the menu.

T F **3.** Professional cooks with experience rely only on the cooking time specified in the standardized recipe and not on the appearance or feel of an item.

T F **4.** Yield may be given as portions, the size of the item produced, or a measured amount of product.

Multiple Choice

________________ **1.** Most standardized recipes usually contain the ___.

A. as-purchased unit cost
B. cooking temperature
C. menu price
D. yield percentage

________________ **2.** While ___ is not required to prepare a recipe, it is important to have for customer inquiries.

A. a preparation procedure
B. an ingredient list
C. the cooking temperature
D. nutrition information

________________ **3.** ___ may include serving and storage instructions.

A. Portion sizes
B. Ingredient lists
C. Preparation steps
D. Nutrition information

Completion

________________ **1.** A(n) ___ is the total quantity of a food or beverage item that is made from a standardized recipe.

______________ **2.** Since the same cook does not always prepare the same items each day, foodservice operations use ___ recipes.

______________ **3.** ___ are usually listed in the order that they are incorporated into a recipe to help ensure that none are left out.

______________ **4.** Differences between the ___ name and the standardized recipe name can lead to confusion in the preparation process.

Matching

______________ **1.** Yield

______________ **2.** Portion size

______________ **3.** Standardized recipe

A. List of ingredients, ingredient amounts, and procedural steps for preparing a specific quantity of a food item

B. Total quantity of a food or beverage item that is made from a standardized recipe

C. Amount of a food or beverage item that is served to an individual person

Chapter 5 Review
Cost Control Fundamentals

Section 5.2 Units of Measure

Name ______________________________ **Date** ______________

True-False

T F **1.** As the count number gets smaller, the size of an item gets larger.

T F **2.** Weight is the measurement of the physical space a substance occupies.

T F **3.** Most food ingredients have a lower density than water and measure more by weight in ounces than by volume in fluid ounces.

T F **4.** The tools used to measure volume are measuring spoons, dry measuring cups, liquid measuring cups, ladles, and portion-controlled scoops.

T F **5.** An ounce is a measurement of weight and a fluid ounce is a measurement of volume.

T F **6.** To convert a customary measurement to a metric measurement, the measurement is multiplied by the number of metric units that equals one of the customary units.

Multiple Choice

______________ **1.** To weigh items on a scale, the empty container that will be used to hold the ingredients is placed on the scale and then the scale is tared by adjusting the scale to read ___.

A. 0
B. 1
C. 2
D. 3

______________ **2.** ___ is the measure of how much a given volume of a substance weighs.

A. Area
B. Mass
C. Density
D. Count

______________ **3.** Converting cups to ounces or teaspoons to grams are examples of converting between ___ measurements.

A. weight and count
B. volume and yield
C. weight and portion
D. volume and weight

______________ **4.** The only customary weight equivalent used in the professional kitchen is ___ oz = 1 lb.
- A. 12
- B. 14
- C. 16
- D. 18

______________ **5.** ___ is the metric unit used to measure weight.
- A. Liters
- B. Grams
- C. Ounces
- D. Pints

______________ **6.** Some ingredients, such as ___, weigh the same amount in ounces as they do in fluid ounces.
- A. water
- B. molasses
- C. corn syrup
- D. honey

Completion

______________ **1.** ___ is the process of changing a measurement with one unit of measure to an equivalent measurement with a different unit of measure.

______________ **2.** When measuring by ___, a measurement can be affected by the technique used to fill the measuring tool.

______________ **3.** When calculating volume equivalents, to change from a larger to a smaller unit of measure, ___ the number in the original measurement by the number of smaller units that make up the larger unit.

______________ **4.** Weight measurement equipment includes mechanical, digital, and balance ___.

______________ **5.** The most common units for measuring ___ in the professional kitchen are the gram (g), ounce (oz), and pound (lb or #).

______________ **6.** To change an ingredient measurement from ounces only to ___ and ounces, the number of ounces is divided by 16.

______________ **7.** ___ is a measurement of the heaviness of a substance.

______________ **8.** ___ is a measurement commonly used for whole ingredients, such as 2 whole eggs or 1 medium banana.

______________ **9.** A(n) ___ is a fixed quantity that is widely accepted as a standard of measurement.

______________ **10.** A(n) ___ is the amount of one unit of measure that is equal to another unit of measure.

Chapter 5 Review
Cost Control Fundamentals

Section 5.3 Scaling Recipes

Name ______________________________ **Date** ________________

True-False

T F **1.** The formula for calculating a scaling factor is as follows: Scaling Factor = Original Yield ÷ Desired Yield.

T F **2.** A standardized recipe produces a specific yield.

T F **3.** When recipes are scaled, instructions related to mixing times never need to be adjusted.

T F **4.** Ingredients can only be scaled by weight.

T F **5.** When scaling recipes, it may be necessary to convert some new ingredient amounts to a different unit of measure.

Multiple Choice

________________ **1.** ___ is the process of calculating new amounts for each ingredient in a recipe when the total amount of food the recipe makes is changed.
- A. Converting
- B. Sizing
- C. Scaling
- D. Portioning

________________ **2.** Once a scaling factor is known, every ingredient amount in the original recipe is ___ the scaling factor.
- A. subtracted from
- B. added to
- C. divided by
- D. multiplied by

________________ **3.** To determine a scaling factor based on ___, divide the available amount (AA) by the original amount (OA).
- A. product availability
- B. portion size
- C. count
- D. volume

Completion

______________________ **1.** A(n) ___ is the number that each ingredient amount in a recipe is multiplied by when the recipe yield is scaled.

______________________ **2.** If a new measurement is calculated that is not easily measured, such as 3.7 cups, it should be adjusted to ___ cups.

______________________ **3.** ___ indicates the heaviness of a substance.

______________________ **4.** The scaling process starts by calculating a(n) ___.

______________________ **5.** ___ indicates the space a substance occupies.

Name ______________________________ Date ____________________

True-False

T F **1.** Unless a food item has a yield percentage of 100%, the edible-portion unit cost will always be lower than the as-purchased unit cost.

T F **2.** A yield percentage is the as-purchased quantity of a food item divided by the edible-portion quantity.

T F **3.** The unit of measure used to calculate a unit cost should be based on how the product is used in recipes.

T F **4.** To calculate the as-served cost of a menu item that is prepared in large quantities and then portioned into servings, the total cost of the recipe is divided by the number of portions the recipe yields.

T F **5.** When calculating the number of servings that can be made from a given recipe, the yield percentage result should be rounded down.

T F **6.** Calculating an edible-portion quantity is done to determine the proper amount of an ingredient to order to make a given quantity of food.

T F **7.** The edible-portion costs of food and beverage products are equal to the prices listed on the invoice.

T F **8.** An as-purchased (AP) quantity is the original amount of a food item as it is ordered and received.

Multiple Choice

______________ **1.** A ___ yield test is a procedure used to determine the yield percentage of a food item that loses weight during the cooking process.

A. raw
B. cooking-loss
C. recipe
D. portion

______________ **2.** The ingredient amounts provided in a recipe are ___ quantities.
A. as-purchased
B. edible-portion
C. yield percentage
D. target percentage

______________ **3.** A(n) ___ is the amount paid for a product in the form it was ordered and received.
A. unit cost
B. edible-portion (EP) quantity
C. as-purchased (AP) cost
D. as-purchased (AP) unit cost

______________ **4.** A ___ test is a procedure used to determine the yield percentage of a food item that is trimmed of waste prior to being used in a recipe.
A. raw yield
B. cooking-loss yield
C. rounding
D. food cost

______________ **5.** When ordering food, the yield percentage results should always be ___.
A. decreased by 10 units of measure
B. increased by 10 units of measure
C. rounded down
D. rounded up

Completion

______________ **1.** An edible-portion quantity is the amount of food item that remains after ___ and is ready to be served or used in a recipe.

______________ **2.** A(n) ___ is the cost of a product per unit of measure.

______________ **3.** A(n) ___ quantity is the original amount of a food item as it is ordered and received.

______________ **4.** A(n) ___ cost is the cost of a menu item as it is served to a customer.

______________ **5.** The percentage circle can be used to help understand how calculations using ___ percentages are performed.

______________ **6.** A(n) ___ test is a procedure used to determine the yield percentage of a food item that is trimmed of waste prior to being used in a recipe.

______________ **7.** A(n) ___ is a document provided by a supplier that lists the items delivered to a foodservice operation and the prices of those items.

______________ **8.** A(n) ___ quantity is the amount of a food item that remains after trimming and is ready to be served or used in a recipe.

Chapter 5 Review
Cost Control Fundamentals

Section 5.5 Food Cost Percentages

Name ______________________________ **Date** ____________________

True-False

T F **1.** The menu price of an item is always higher than the as-served cost of that item.

T F **2.** A menu-item food cost percentage is equal to the menu price of a menu item divided by the AS cost written as a percent.

T F **3.** Lower food cost percentages translate into higher potential earnings for the owners of the operation.

T F **4.** An overall beverage cost percentage is a percentage that indicates how the cost of beverages relates to menu prices and beverage sales of a foodservice operation.

Multiple Choice

______________ **1.** A(n) ___ food cost percentage is the AS cost of a menu item divided by the menu price and written as a percent.

A. edible-portion
B. overall
C. menu-item
D. target

______________ **2.** If the menu price for a chicken sandwich is $6.95 and the AS cost is $1.80, the menu-item food cost percentage for the chicken sandwich would be ___%.

A. 14.6
B. 25.9
C. 36.2
D. 41.5

______________ **3.** If a restaurant spends $9000 on food in a month when its total food sales were $36,000 for that month, the overall food cost percentage for the restaurant would be ___%.

A. 10
B. 15
C. 20
D. 25

______________ **4.** When the overall food cost percentage of a foodservice operation is lower than the target food cost percentage, the operation has ___.
A. earned more money selling food than planned
B. lost inventory to spoilage and trimming
C. increased spending on food and beverages
D. broken even

Completion

______________ **1.** A(n) ___ food cost percentage is the total amount of money a foodservice operation spends on food divided by the total food sales over a defined period of time.

______________ **2.** A(n) ___ food cost percentage is the percentage of food sales that a foodservice operation plans to spend on purchasing food.

______________ **3.** A(n) ___ is a percentage that indicates how the cost of food relates to the menu prices and food sales of a foodservice operation.

______________ **4.** If the actual overall food cost percentage is ___ than the target food cost percentage, the operation has earned less money selling food than planned.

Name ______________________________ **Date** ______________

True-False

T F **1.** A pricing form is a tool often used to help calculate the as-served cost of a menu item and establish a menu price.

T F **2.** After pricing forms have been completed for all of the items on a menu, the pricing forms never need readjusting.

T F **3.** It is normal for items on a menu to have different food cost percentages.

T F **4.** If the target price of a bowl of soup is calculated to be $4.67, the restaurant might increase the price to $4.95.

Multiple Choice

______________ **1.** A(n) ___ price is the price that a foodservice operation needs to charge for a menu item in order to meet its target food cost percentage.
- A. overall
- B. perceived
- C. ideal
- D. target

______________ **2.** ___ pricing is not common in restaurants but is often used when calculating a price charged per person for a special event.
- A. Contribution margin
- B. Perceived value
- C. As-served cost
- D. Food cost percentage

______________ **3.** The ___ is calculated by dividing the total cost by the number of portions.
- A. target price
- B. as-served cost per portion
- C. yield percentage
- D. edible-portion quantity

______________ **4.** When completing a pricing form, the ___ is calculated by adding the costs of each ingredient.
- A. menu price
- B. as-purchased (AP) unit cost
- C. total cost
- D. as-served (AS) cost per portion

Completion

______________ **1.** Perceived value pricing is the process of adjusting a target menu price based on how management thinks a(n) ___ will perceive the price.

______________ **2.** A(n) ___ is the amount added to the AS cost of a menu item to determine a menu price.

______________ **3.** A(n) ___ is the price listed on a menu for a single portion of the menu item.

______________ **4.** The ___ is the unit cost for each ingredient in the same unit of measure as the EP quantity of each ingredient.

Name ______________________________ Date ________________

True-False

T F **1.** Food and beverage products and payroll expenses are examples of fixed costs.

T F **2.** When planning a profitable menu, seasonal ingredients do not need to be taken into account.

T F **3.** Perishable foods should be purchased frequently in the smallest quantities possible.

T F **4.** Rent, real estate taxes, and insurance are examples of fixed costs.

Multiple Choice

______________ **1.** A(n) ___ is a written form listing the specific characteristics of a product that is to be purchased from a supplier.
A. invoice
B. requisition
C. purchase specification
D. pricing form

______________ **2.** A(n) ___ is a document that markets the foodservice operation to the customer.
A. menu
B. requistion
C. invoice
D. purchase specification

______________ **3.** Nonperishable food items generally can be kept for ___.
A. 3 to 4 months
B. 6 months to a year
C. 1 to 2 years
D. up to 36 months

______________ **4.** All food and beverage items in dry and cold storage areas must be kept ___ inches off the floor.
A. 2
B. 4
C. 6
D. 10

Completion

______________ **1.** Par stock is the ___ amount of a particular product that should be kept in inventory to ensure that an adequate supply is on hand for normal production.

______________ **2.** A(n) ___ is an internally generated invoice that is used to aid in tracking inventory as it moves from storage to production.

______________ **3.** ___ is the process of ensuring that a specific amount of food or beverage is served for a given price.

______________ **4.** A(n) ___ is the amount of money earned by an operation when revenue is greater than expenses.

Matching

______________ **1.** Variable cost

______________ **2.** Fixed cost

______________ **3.** Par stock

______________ **4.** Purchase specification

A. Cost that does not change as sales increase or decrease

B. Maximum amount of a particular product that should be kept in inventory to ensure that an adequate supply is on hand for normal production

C. Written form listing the specific characteristics of a product to be purchased from a supplier

D. Cost that increases or decreases in proportion to the volume of production

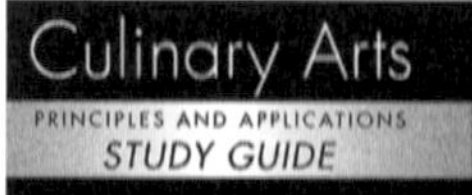

Chapter 5 Review
Cost Control Fundamentals

Section 5.8 Standard Profit and Loss

Name ______________________ **Date** ______________

True-False

T F **1.** Payroll expenses are frequently the most significant expense of a foodservice operation.

T F **2.** Foodservice operations are required to carry various types of insurance such as liability insurance and workers compensation insurance.

T F **3.** A loss is uncommon during the first several months of a new operation.

T F **4.** The gross pay for a salaried employee is not the same for every pay period.

T F **5.** The payroll expenses of salaried employees are considered fixed expenses.

Multiple Choice

______________ **1.** Net profit is the calculated difference between the ___ profit and operating expenses of a foodservice operation.

A. total
B. gross
C. fixed
D. variable

______________ **2.** The ___ profit is the last entry on the standard profit and loss statement and is referred to as the "bottom line."

A. variable
B. fixed
C. gross
D. net

______________ **3.** Hourly employees are required to be paid at least ___ times their regular hourly wage for all hours over 40 worked in a week.

A. 1.5
B. 2
C. 2.5
D. 3

______ **4.** The purpose of a ___ is to provide management with an indication of the financial status of a foodservice operation.
- A. monthly bill
- B. pricing form
- C. standard profit and loss statement
- D. monthly earnings graph

______ **5.** The net pay of each employee's paycheck will be less than the employee's ___.
- A. taxes
- B. gross pay
- C. net profit
- D. gross profit

______ **6.** The gross pay for a line cook who earns $12 per hour and works 50 hours in one week would equal about $___.
- A. 480
- B. 510
- C. 560
- D. 660

Completion

______ **1.** A standard ___ statement is a form that shows the revenue, expenses, and resulting gross and net profit (or loss) over a specific period of time.

______ **2.** ___ pay is the total amount of an employee's pay before any deductions are made.

______ **3.** If a restaurant with a gross profit of $20,500 had $18,250 in operating expenses, there would be a net profit of $___.

______ **4.** A(n) ___ employee is paid based on a fixed amount of money.

______ **5.** Overtime pay is equal to any hours worked over ___ hours in a week multiplied by the overtime hourly wage.

______ **6.** ___ is the calculated difference between total revenue and the cost of goods sold.

Chapter 5 Activities
Cost Control Fundamentals

Name ______________________ Date ______________

Activity: Determining Measurement Equivalents

A measurement equivalent is the amount of one unit of measure that is equal to another unit of measure. Foodservice employees need to know the basic measurement equivalents for both weight and volume and be able to use these equivalents to convert between units.

Complete the following using the appropriate measuring equipment.

__________ **1.** A dry measure of ___ tsp is equivalent to 1 tbsp.

__________ **2.** A dry measure of 2 tbsp is equivalent to ___ cup(s).

__________ **3.** A dry measure of 4 tbsp is equivalent to ___ cup(s).

__________ **4.** A dry measure of 5⅓ tbsp is equivalent to ___ cup(s).

__________ **5.** A dry measure of 8 tbsp is equivalent to ___ cup(s).

__________ **6.** A dry measure of ___ tbsp is equivalent to 1 cup.

__________ **7.** A liquid measure of 2 tbsp is equivalent to ___ fl oz.

__________ **8.** A liquid measure of ___ tbsp is equivalent to 2 fl oz.

__________ **9.** A liquid measure of 8 tbsp is equivalent to ___ fl oz.

__________ **10.** A liquid measure of 1 cup is equivalent to ___ fl oz.

__________ **11.** A liquid measure of ___ cup(s) is equivalent to 1 pt.

__________ **12.** A liquid measure of 1 pt is equivalent to ___ fl oz.

__________ **13.** A liquid measure of ___ cup(s) is equivalent to 1 qt.

__________ **14.** A liquid measure of 1 qt is equivalent to ___ fl oz.

__________ **15.** A liquid measure of 8 cups is equivalent to ___ fl oz.

__________ **16.** A liquid measure of 16 cups is equivalent to ___ pt.

__________ **17.** A liquid measure of ___ pt is equivalent to 1 gal.

__________ **18.** A liquid measure of ___ qt is equivalent to 1 gal.

__________ **19.** A liquid measure of 1 gal. is equivalent to ___ fl oz.

__________ **20.** A liquid measure of 1 pt is equivalent to ___ qt.

Activity: Scaling and Standardizing Recipes

After scaling ingredients for a recipe, the recipe can be standardized. Standardizing recipes is the key to having menu items turn out the same way consistently. To scale a recipe, use the following formula:

$SF = DY \div OY$

where

SF = scaling factor

DY = desired yield

OY = original yield

Scale the following recipes to yield 100 servings and organize the recipes into standardized form.

1. Scale the following tomato cream cheese bouchées recipe to yield 100 servings.

Tomato Cream Cheese Bouchées (Yield: 4 servings, 4 each)		
Original Yield ×	**SF** =	**Desired Yield**
2 tbsp olive oil		
1 clove minced garlic		
1 ea medium tomato		
½ tbsp fresh basil		
TT salt and pepper		
8 oz softened cream cheese		
½ lb puff pastry dough		

2. Organize the recipe into standardized form.

Tomato Cream Cheese Bouchées	
Yield: 100 servings **Portion Size:** 4 each	**Cooking Temperature:** 375°F **Cooking Time:** 10 –12 min
Ingredients	**Procedure**

3. Scale the grilled chicken breast with sweet teriyaki marinade recipe to yield 100 servings.

Grilled Chicken Breast with Sweet Teriyaki Marinade (Yield: 4 servings, 2 breasts each)		
Original Yield ×	SF =	Desired Yield
4 oz soy sauce		
½ c mirin or sweet cherry		
2 tbsp rice wine vinegar		
2 tbsp vegetable oil		
2 oz granulated sugar		
2 tbsp grated or minced fresh ginger		
2 oz lager-style beer		
8 ea boneless chicken breasts		

4. Organize the recipe into standardized form.

Grilled Chicken Breast with Sweet Teriyaki Marinade **Yield:** 100 servings **Portion Size:** 2 breasts each	**Cooking Temperature:** N/A **Cooking Time:** N/A
Ingredients	**Procedure**

5. Scale the sabayon recipe to yield 100 servings.

Sabayon (Yield: 10 servings, 2 oz each)		
Original Yield ×	SF =	Desired Yield
8 oz champagne		
10 egg yolks		
pinch salt		
8 oz granulated sugar		

6. Organize the recipe into standardized form.

Sabayon	
Yield: 100 servings **Portion Size:** 2 oz each	**Cooking Temperature:** N/A **Cooking Time:** N/A
Ingredients	**Procedure**

Activity: Calculating As-Purchased Unit Costs

The as-purchased (AP) unit cost is the cost of the item per unit. The AP unit cost (APU) is found by applying the following formula:

$APU = APC \div NU$

where

APU = AP unit cost

APC = AP cost

NU = number of units

Calculate the AP unit cost for each of the following items.

______ **1.** The APU of milk is $___ per cup when 4 gal. of milk is purchased for $13.88.

______ **2.** The APU of eggs is $___ per egg when 15 dozen eggs are purchased for $17.67.

______ **3.** The APU of red potatoes, size B, is $___ per lb when 50 lb are purchased for $24.00.

______ **4.** The APU of raspberries is $___ per ½ pt when 6 pt are purchased for $54.00.

______ **5.** The APU of fresh basil is $___ per oz when 1 lb is purchased for $11.26.

______ **6.** The APU of beef stew meat is $___ per lb when 10 lb are purchased for $25.21.

______ **7.** The APU of boneless, skinless chicken breast is $___ per lb when 20 lb are purchased for $52.01.

______ **8.** The APU of sliced bacon is $___ per slice when a case of twenty 16-slice packages is purchased for $55.88.

______ **9.** The APU of littleneck clams is $___ per clam when 25 clams are purchased for $7.99.

______ **10.** The APU of ciabatta bread is $___ per loaf when 25 loaves are purchased for $42.50.

______ **11.** The APU of all-purpose flour is $___ per lb when a 50 lb bag is purchased for $60.00.

______ **12.** The APU of macaroni is $___ per oz when a 30 lb package is purchased for $24.41.

______ **13.** The APU of spring mix lettuce is $___ per lb when a 3 lb container is purchased for $24.25.

Activity: Calculating Edible-Portion Unit Costs

The edible-portion (EP) unit cost is the unit cost of a food or beverage after taking into account the cost of the waste generated by trimming. The EP unit cost (EPU) is found by applying the following formula:

$EPU = APU \div YP$

where

EPU = EP unit cost

APU = AP unit cost

YP = yield percentage

Calculate the EP unit cost of the following items.

______________ **1.** The EPU of broccoli is $___ when the APU is $1.09 and the YP is 70%.

______________ **2.** The EPU of parsley is $___ when the APU is $0.39 and the EP is 76%.

______________ **3.** The EPU of shallots is $___ when the APU is $11.50 and the YP is 89%.

______________ **4.** The EPU of carrots is $___ when the APU is $17.00 and the YP is 80%.

______________ **5.** The EPU of asparagus is $___ when the APU is $32.67 and the YP is 56%.

______________ **6.** The EPU of green beans is $___ when the APU is $28.98 and the YP is 88%.

______________ **7.** The EPU of winter squash is $___ when the APU is $2.50 and the YP is 68%.

______________ **8.** The EPU of leeks is $___ when the APU is $1.54 and the YP is 52%.

______________ **9.** The EPU of Brussels sprouts is $___ when the APU is $2.50 and the YP is 74%.

______________ **10.** The EPU of zucchini is $___ when the APU is $2.39 and the YP is 92%.

Activity: Calculating Recipe Costs

Calculating food costs involves determining the total cost of a recipe based on the individual costs of all ingredients. When calculating total recipe cost, it is important to figure the cost for each item using the same unit of measure that is called for in the recipe.

Calculate the total cost of the shrimp creole recipe using the following form.

Recipe Costing Form

Recipe: Shrimp Creole

Yield: 25 Servings

Portion Size: 9 oz

Item No.	Ingredient	Quantity	As-Purchased (AP) Amount	As-Purchased (AP) Cost	Yield Percentage	Edible-Portion (EP) Amount	Edible-Portion (EP) Cost	Total Cost
1532	**8 lb shrimp**	8 lb	10 lb	$119.50	88%			
254	**3 qt prepared creole sauce**	3 qt	—	—	—	—	—	—
1906	**olive oil**	¼ c	1 gal.	$25.86	100%			
1148	**medium onion, chopped**	6 onions	50 lb	$75.00	83%			
1112	**celery, chopped**	3 c	25 lb	$28.75	85%			
1126	**red bell pepper, chopped**	3 peppers	25 lb	$35.99	85%			
1159	**garlic, minced**	6 cloves	50 cloves	$53.99	81%			
103	**chicken stock**	1.5 qt	1 gal.	$1.03	100%			
2106	**canned tomatoes, chopped**	5.5 lb	6.83 lb	$5.26	100%			
1874	**bay leaf**	3 leaves	8 oz	$5.75	100%			
1852	**Cajun seasoning**	2 oz	20 oz	$4.17	100%			
1886	**salt**	TT	25 lb	$30.00	100%			
1890	**pepper**	TT	5 lb	$40.00	100%			
1216	**prepared white rice**	3 lb	25 lb	$14.25	100%			
							Total	

_______________ **1.** The EP amount of shrimp is ___.

_______________ **2.** The EP amount of olive oil is ___.

_______________ **3.** The EP amount of onion is ___.

_______________ **4.** The EP amount of celery is ___.

_______________ **5.** The EP amount of red bell pepper is ___.

_______________ **6.** The EP amount of garlic is ___.

_______________ **7.** The EP amount of chicken stock is ___.

_______________ **8.** The EP amount of canned tomatoes is ___.

_______________ **9.** The EP amount of bay leaf is ___.

_______________ **10.** The EP amount of Cajun seasoning is ___.

_______________ **11.** The EP amount of salt is ___.

_______________ **12.** The EP amount of pepper is ___.

_______________ **13.** The EP amount of prepared white rice is ___.

_______________ **14.** The EP cost of shrimp is $___ per lb.

_______________ **15.** The EP cost of olive oil is $___ per gal.

_______________ **16.** The EP cost of onion is $___ per lb.

_______________ **17.** The EP cost of celery is $___ per lb.

_______________ **18.** The EP cost of red bell pepper is $___ per lb.

_______________ **19.** The EP cost of garlic is $___ per clove.

_______________ **20.** The EP cost of chicken stock is $___ per gal.

_______________ **21.** The EP cost of canned tomatoes is $___ per lb.

_______________ **22.** The EP cost of bay leaf is $___ per oz.

_______________ **23.** The EP cost of Cajun seasoning is $___ per oz.

_______________ **24.** The EP cost of salt is $___ per lb.

_______________ **25.** The EP cost of pepper is $___ per lb.

_______________ **26.** The EP cost of prepared white rice is $___ per lb.

_______________ **27.** The total cost of shrimp is $___.

_______________ **28.** If there are 16 cups in a gal., the total cost of olive oil is $___.

_______________ **29.** If an onion weighs ⅓ lb, the total cost of onion is $___.

_______________ **30.** If 1 cup of celery weighs ⅓ lb, the total cost of celery is $___.

__________ **31.** If a red bell pepper weighs 4 oz, the total cost of red bell pepper is $___.

__________ **32.** The total cost of garlic is $___.

__________ **33.** If there are 4 qt in a gal., the total cost of chicken stock is $___.

__________ **34.** The total cost of tomatoes is $___.

__________ **35.** If a bay leaf weighs 0.5 oz, the total cost of bay leaf is $___.

__________ **36.** The total cost of Cajun seasoning is $___.

__________ **37.** If ⅕ oz of salt is used to season to taste, the total cost of salt is $___.

__________ **38.** If ⅕ oz of pepper is used to season to taste, the total cost of pepper is $___.

__________ **39.** The total cost of prepared white rice is $___.

__________ **40.** The total cost for the shrimp creole recipe is $___.

__________ **41.** The cost per serving for the Shrimp Creole recipe is $___.

Activity: Calculating Food Cost Percentages

A food cost percentage is a percentage that indicates how the cost of food relates to the menu prices and food sales of a foodservice operation. A target food cost percentage is the percentage of total food sales that a foodservice operation plans to spend on purchasing food. Food cost percentages are compared with target food cost percentages to determine if food costs are higher or lower than planned.

Complete the items using the following Daily Sales Report.

Daily Sales Report—Lunch

Target Food Cost Percentage = 30%

Menu Item	AS Cost	Menu Price	Menu Item Food Cost % (AS Cost ÷ Menu Price)	Number Sold	Total Food Cost (AS Cost × Number Sold)	Total Food Sales (Menu Price × Number Sold)
Appetizers						
Bruschetta	$1.75	$5.95	29.4%	48		
BBQ shrimp	$4.25	$11.95	35.6%	27		
Entrées						
Fried chicken	$2.65	$12.95	20.5%	43		
Strip steak	$9.00	$25.95	34.7%	32		
Desserts						
Strawberry shortcake	$2.25	$5.95	37.8%	40		
Chocolate cake	$0.50	$3.95	12.7%	30		
Total						
Overall Total Food Cost Percentage (Total Food Costs ÷ Total Food Sales)						

______________________ **1.** The total food cost for the bruschetta appetizer is $___.

______________________ **2.** The total food cost for the BBQ shrimp appetizer is $___.

______________________ **3.** The total food cost for the fried chicken entrée is $___.

______________________ **4.** The total food cost for the strip steak entrée is $___.

______________________ **5.** The total food cost for the strawberry shortcake dessert is $___.

______________________ **6.** The total food cost for the chocolate cake dessert is $___.

______________________ **7.** The total food cost is $___.

______________________ **8.** The total food sales for the bruschetta appetizer are $___.

______________________ **9.** The total food sales for the BBQ shrimp appetizer are $___.

______________________ **10.** The total food sales for the fried chicken entrée are $___.

______________________ **11.** The total food sales for the strip steak entrée are $___.

______________________ **12.** The total food sales for the strawberry shortcake dessert are $___.

_______________ **13.** The total food sales for the chocolate cake dessert are $___.

_______________ **14.** The total food sales are $___.

_______________ **15.** The overall food cost percentage is ___%.

16. What does it indicate if the overall food cost percentage is lower than the target food cost percentage?

17. What does it indicate if the overall food cost percentage is higher than the target food cost percentage?

Activity: Calculating Net Profits

A standard profit and loss statement is a form that shows the revenue, expenses, and resulting gross and net profit (or loss) over a specific period of time. The purpose of a standard profit and loss statement is to provide management with an indication of the financial status of the foodservice operation.

Complete the following questions using the following Standard Profit and Loss Statement.

1. What is the total revenue of the foodservice operation?

2. What is the cost of goods sold for the foodservice operation?

3. What is the gross profit for the foodservice operation?

4. What are the total operating expenses for the foodservice operation?

5. What is the net profit (or loss) for the foodservice operation?

6. What does the net profit (or loss) indicate to management about this foodservice operation?

Standard Profit and Loss Statement Detailed		
Revenue		
Total food revenue	$	27,801.23
Total beverage revenue	$	7107.36
Total revenue	**$**	
Cost of goods sold		
Total food costs	$	7794.38
Total beverage costs	$	2638.52
Total cost of goods sold	**$**	
Gross profit (Total revenue – Total cost of goods sold)	**$**	
Operating expenses		
Payroll expenses	$	13,010.15
Social security taxes	$	1050.75
Utilities	$	1124.67
Rent	$	2800.00
Interest	$	325.19
Insurance and licenses	$	356.80
Kitchen supplies	$	803.21
Sales tax	$	1047.26
Miscellaneous	$	250.22
Total operating expenses	**$**	
Net profit / loss (Gross profit – Total operating expenses)	**$**	

Chapter 6 Review
Menu Planning and Nutrition

Section 6.1 Menu Styles

Name ______________________________ **Date** ________________

True-False

T F **1.** In general, menu items that are common or familiar do not require descriptions.

T F **2.** A cycle menu is a menu that changes frequently to coincide with changes in the availability of products.

T F **3.** The eye naturally falls on the top left of a single-page menu.

T F **4.** A menu is a list of items that guests may order from a foodservice operation.

T F **5.** Multipage menus have a right and a left panel that fold toward the center and overlap.

Multiple Choice

________________ **1.** All types of menus, except the ___ menu, can be meal specific.
A. à la carte
B. semi-à la carte
C. prix fixe
D. California

________________ **2.** A ___ menu is a menu that offers limited choices within a collection of specific items for a multicourse meal at a set price.
A. table d'hôte
B. prix fixe
C. California
D. meal-specific

________________ **3.** The Patient Protection and Affordable Care Act of 2010 requires foodservice operations with 20 or more locations to ___ for standard menu items.
A. provide kid meals
B. offer healthy alternatives
C. list the calories
D. offer vegan options

______ **4.** A ___ menu is a menu that identifies specific items that will be served for each course at a set price.
A. table d'hôte
B. prix fixe
C. semi-à la carte
D. meal-specific

______ **5.** ___ menus present menu items on one page and may use both the front and back of the sheet.
A. Bifold
B. Multipage
C. Trifold
D. Single-page

Completion

______ **1.** A(n) ___ is the primary form of communication that a foodservice operation uses to inform guests of the types of foods and beverages offered and how much each item costs.

______ **2.** A(n) ___ menu is a menu that offers entrées along with additional menu items for a set price.

______ **3.** The ___ guidelines are designed to protect guests from fraudulent food and beverage claims.

______ **4.** ___ menus are commonly used by restaurants that are open 24 hours a day.

______ **5.** A(n) ___ menu is a menu that only offers a particular meal, such as breakfast, lunch, or dinner.

Matching

______ **1.** Fixed menu

______ **2.** Cycle menu

______ **3.** Menu mix

______ **4.** Market menu

A. A menu that changes frequently to coincide with changes in the availability of products

B. A menu that stays the same or rarely changes

C. The assortment of items that may be ordered from a given menu

D. A menu written for a specific period and repeated once that period ends

Name ______________________________ **Date** ____________________

True-False

T F **1.** With effective pricing, a menu item can be more profitable even though it has a higher food cost percentage.

T F **2.** Finding the right balance of menu prices is a complex process influenced by costs, perceived value, and the placement of prices on the menu.

T F **3.** If menu prices are listed from least expensive to most expensive or vice versa, guests will be encouraged to read through all the menu offerings.

Multiple Choice

________________ **1.** Foodservice operations generate ___ when they charge more for the items they sell than the amount spent to produce those items.
A. costs
B. losses
C. profits
D. debt

________________ **2.** ___ represents the amount that guests are willing to pay for a menu item.
A. Menu price
B. Perceived value
C. Cost view
D. Market value

________________ **3.** A menu item with a target price of $4.47 would be rounded to $___.
A. 4.00
B. 4.45
C. 4.50
D. 5.00

Completion

______________ 1. Ideally, effectively priced menu items will generate the greatest ___ while being perceived as a good value by the guests.

______________ 2. Before a menu price can be established, a foodservice operation must determine the ___ of each menu item.

______________ 3. If a pasta dish sells for $10.00 with a 25% food cost percentage, the operation makes a $___ profit.

Chapter 6 Review
Menu Planning and Nutrition

Section 6.3 Menu Dietary Considerations

Name ______________________________ **Date** ____________________

True-False

T F **1.** Anaphylaxis is a severe allergic reaction that causes the airway to narrow and prohibits breathing.

T F **2.** Symptoms of food intolerance usually include bloating, abdominal cramps, nausea, and diarrhea.

T F **3.** Celiac disease is an inability to properly digest lactose.

T F **4.** All allergic reactions are extremely severe and life-threatening.

T F **5.** Gluten is prevalent in cereals, pastas, and baked goods.

T F **6.** There are less than 50 foods that can cause allergic reactions.

Multiple Choice

__________ **1.** The rise in ___ has paralleled an increase in illnesses and diseases, such as high blood pressure, cardiovascular disease, type 2 diabetes, and some cancers.
A. obesity
B. nut allergies
C. lactose intolerance
D. Celiac disease

__________ **2.** Lactose is a ___ found in milk and dairy products.
A. vitamin
B. mineral
C. starch
D. sugar

__________ **3.** ___-based meals are also known as vegetarian meals.
A. Herb
B. Plant
C. Seafood
D. Egg

_______________ **4.** Gluten is a type of protein found in ___.
A. dairy products
B. fats and oils
C. grains
D. nuts

_______________ **5.** According to the U.S. Department of Health and Human Services, the number of individuals who are overweight or obese has risen from 13% to ___% over the past several decades.
A. 22
B. 35
C. 48
D. 67

_______________ **6.** Vegetarian menus may include ___ for individuals looking to exclude animal-based foods.
A. chickpea coconut curry stew
B. veal parmesan
C. gyros
D. quiche lorraine

Completion

_______________ **1.** According to the FDA, milk, eggs, fish, crustacean shellfish, tree nuts, peanuts, soybeans, and ___ account for 90% of all food allergies.

_______________ **2.** A food intolerance is an adverse reaction to a food that does not involve the ___ system.

_______________ **3.** Celiac disease is a condition in which ___ damages the small intestine's ability to absorb nutrients.

_______________ **4.** To combat obesity, federal initiatives have been developed to promote healthy eating patterns and ___ activity.

_______________ **5.** The most common food intolerances are gluten intolerance and ___ intolerance.

Matching

_______________ **1.** Obesity

_______________ **2.** Food allergy

_______________ **3.** Nutrient-dense food

A. A food that is high in nutrients and low in calories

B. A medical condition characterized by an excess of body fat

C. A reaction by the immune system to a specific food

Chapter 6 Review
Menu Planning and Nutrition

Section 6.4 Nutrition Fundamentals

Name ______________________________ **Date** ____________________

True-False

T F **1.** An unsaturated fat is a lipid that is solid at room temperature.

T F **2.** In addition to eggs, complete proteins are primarily found in animal-based foods such as meat, poultry, fish, and dairy.

T F **3.** A complex carbohydrate is a carbohydrate composed of one or two sugar units and is quickly absorbed by the body.

T F **4.** Nearly every body function is dependent upon water.

T F **5.** Dietary fiber is the portion of a plant that the body cannot digest.

Multiple Choice

______________ **1.** The primary function of ___ is to build and repair body tissue, but the body can also convert it to energy when needed.

A. carbohydrates
B. protein
C. water
D. lipids

______________ **2.** Water-soluble vitamins include vitamin ___ and the vitamin B complex.

A. C
B. D
C. E
D. K

______________ **3.** ___ need to be replenished daily because they are not stored by the body.

A. Complete proteins
B. Incomplete proteins
C. Water-soluble vitamins
D. Fat-soluble vitamins

____________ **4.** Both ___ contribute to increased levels of cholesterol.
A. trans fats and nonessential amino acids
B. dietary fiber and saturated fats
C. saturated fats and trans fats
D. unsaturated fats and saturated fats

____________ **5.** ___ is a process that chemically transforms oils into solids to improve shelf life and stabilize flavor.
A. Anaphylaxis
B. Saturation
C. Digestion
D. Hydrogenation

Completion

____________ **1.** A(n) ___ amino acid is an amino acid that can be made by the body in sufficient quantities.

____________ **2.** ___ help the body absorb nutrients, provide insulation, manufacture hormones, and cushion organs.

____________ **3.** ___-soluble vitamins include vitamins A, D, E, and K.

____________ **4.** ___ fats are often referred to as good fats because they have been found to lower the risk of cardiovascular disease and stroke.

____________ **5.** A nutrient-dense food is a food that is high in nutrients and low in ___.

Matching

____________ **1.** Carbohydrate

____________ **2.** Mineral

____________ **3.** Vitamin

____________ **4.** Lipid

____________ **5.** Protein

A. A nutrient in the form of a fat, oil, or fatty acid

B. A nutrient that consists of one or more chains of amino acids and is essential to living cells

C. An inorganic substance that is required in very small amounts to help regulate body processes

D. A nutrient in the form of sugar or starch that is the human body's main source of energy

E. A nutrient that is composed of organic substances and required in small amounts to help regulate body processes

Chapter 6 Review
Menu Planning and Nutrition

Section 6.5 Dietary Recommendations

Name ______________________________ **Date** ____________________

True-False

T F **1.** The USDA created the ChooseMyPlate website to help people understand how to implement dietary guidelines and develop a healthy eating pattern.

T F **2.** The RDA of grains is 3–8 oz, regardless of the age, gender, and level of physical activity of the person.

T F **3.** Choosing leaner cuts of meat and eating at least 8 oz of cooked seafood per week can help reduce saturated fat and calorie intake.

T F **4.** Most vegetables are naturally low in fat and calories but contain cholesterol.

T F **5.** Regular physical activity can reduce stress and increase the amount of calories used by the body.

T F **6.** The first section of a nutrition facts label includes the serving size and the number of servings per container.

T F **7.** Refined grains maintain vitamins, minerals, and dietary fiber during processing.

T F **8.** The RDA of fruits ranges from 1–2 cups depending on the age, gender, and level of physical activity of the person.

Multiple Choice

______________ **1.** The RDA of vegetables ranges from 1–___ cups depending on the age, gender, and level of physical activity of the person.
A. 2
B. 3
C. 4
D. 5

______________ **2.** The RDA of ___ is 2–6½ oz, depending on the age, gender, and level of physical activity of the person.
A. carbohydrates
B. fruits
C. dairy foods
D. protein foods

_______________ **3.** The RDA of dairy foods is 2–___ cups per day, depending on the age, gender, and level of physical activity of the person.
A. 3
B. 4
C. 5
D. 6

_______________ **4.** The ___ is the largest component of a grain kernel and consists of carbohydrates and a small amount of protein.
A. germ
B. bran
C. endosperm
D. husk

_______________ **5.** ___ is broken down into dietary fiber and sugars on nutrition facts labels.
A. Protein
B. Total carbohydrate
C. Fat
D. Cholesterol

Completion

_______________ **1.** Foods made from milk that have little to no ___, such as butter, cream, and cream cheese, are not considered dairy foods.

_______________ **2.** A whole grain is a grain that only has the ___ removed.

_______________ **3.** Every five years, the U.S. Department of Agriculture (USDA) publishes the ___, which provides evidence-based recommendations about the components of a healthy and nutritionally adequate diet.

_______________ **4.** A(n) ___ is a unit of measurement that represents the amount of energy in a food.

_______________ **5.** A(n) ___ grain is a refined grain that has thiamine, riboflavin, niacin, folate, and iron added to it.

Chapter 6 Review
Menu Planning and Nutrition

Section 6.6 Recipe Modifications

Name ______________________________ **Date** ____________________

True-False

T F **1.** Olive oil can replace a saturated fat, such as butter, to reduce the fat content of a dish.

T F **2.** Substituting some or all the meat in a recipe with vegetables or whole grains can decrease the amount of fat.

T F **3.** By using smaller plates and appropriate portions, menu items can offer guests healthful options that meet dietary guideline recommendations.

Multiple Choice

____________________ **1.** Sugar content can be reduced by using herbs and spices such as ___ to naturally enhance sweetness.

A. mint
B. chipotle
C. cayenne
D. dill

____________________ **2.** Instead of salt, flavor can be enhanced with peppers, herbs, spices, or a(n) ___, such as lemon juice or vinegar.

A. alkali
B. fat
C. sugar
D. acid

____________________ **3.** Menus that offer half of a roasted Cornish hen instead of a whole roasted Cornish hen help promote ___.

A. paleo
B. obesity
C. portion control
D. vegetarianism

Completion

______________________ **1.** ___ software helps the user easily substitute ingredients and change quantities to create healthier recipes.

______________________ **2.** Cooking foods in oil or butter results in higher-___ menu items.

______________________ **3.** The majority of ___ consumed by the average person comes from processed foods.

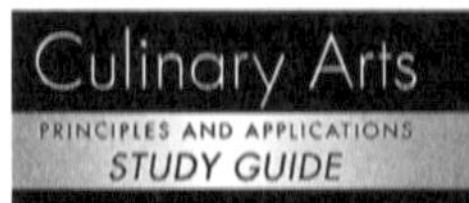

Name ______________________________ Date ________________

Activity: Analyzing Menu Types

Planning menus involves determining the most appropriate menu type for the operation. Menu types are influenced by the menu classification as well as the nature of the foodservice operation.

Collect one example of each of the following menu types.

- À la carte
- Semi-à la carte
- Prix fixe
- Table d'hôte
- California
- Meal-specific

1. Create a report describing the effect menu classification has on the design and pricing for each menu collected.

Activity: Creating Menu Descriptions

Menu descriptions can entice guests by creating tempting images of menu items in their minds. Some words have more selling power than others and can be used to help direct guests to higher-profit items. Enticing menu descriptions promote clarity, add appeal to the dish, and increase the likelihood that an item will be ordered. The following example illustrates how descriptive words can add interest to a dish and tempt the palette.

- Original menu description

  Stuffed Pork Tenderloin – Pork tenderloin stuffed with sausage and apples and served with BBQ sauce

- Enticing menu description

  Stuffed Pork Tenderloin – Hand-trimmed pork tenderloin filled with a flavorful blend of sweet Italian sausage and Granny Smith apples, pan-seared, and glazed with a mandarin orange and honey BBQ sauce

Create an enticing menu description for each of the following items commonly found on menus. Include any sides or special components of each dish.

1. Buffalo wings

Enticing menu description:

2. Chicken noodle soup

Enticing menu description:

3. Caesar salad

Enticing menu description:

4. Cheeseburger

Enticing menu description:

5. Grilled chicken breast

Enticing menu description:

6. Pecan pie

Enticing menu description:

Activity: Creating Meal Plans

Every five years, the USDA publishes *Dietary Guidelines for Americans* that include key recommendations for healthy eating and physical activity. These recommendations stress the importance of the daily consumption of vegetables and fruits, whole grains, and fat-free or low-fat milk and dairy products. These recommendations also place emphasis on consuming protein from a variety of sources, including lean meats, poultry, fish, eggs, soy and nuts, and reducing saturated fats, cholesterol, sodium, and sugar.

Create a three day meal plan that meets the recommended daily allowances for vegetables and fruits, grains, protein foods, and dairy foods based on a 2000 calorie diet.

1. What will be served on day 1?

Breakfast:

Lunch:

Dinner:

Snacks:

2. What will the total number of calories consumed on day 1 be?

3. Is the total number of calories consumed on day 1 more or less than the 2000 calorie recommended daily allowance?

4. Do the meals planned for day 1 provide 1–3 cups of vegetables, 1–2 cups of fruit, 3–8 oz of grains with half from whole grains, 2–6½ oz of protein foods, and 2–3 cups of dairy foods?

5. What will be served on day 2?

Breakfast:

Lunch:

Dinner:

Snacks:

6. What will the total number of calories consumed on day 2 be?

7. Is the total number of calories consumed on day 2 more or less than the 2000 calorie recommended daily allowance?

8. Do the meals planned for day 2 provide 1–3 cups of vegetables, 1–2 cups of fruit, 3–8 oz of grains with half from whole grains, 2–6½ oz of protein foods, and 2–3 cups of dairy foods?

9. What will be served on day 3?

Breakfast:

Lunch:

Dinner:

Snacks:

10. What will the total number of calories consumed on day 3 be?

11. Is the total number of calories consumed on day 3 more or less than the 2000 calorie recommended daily allowance?

12. Do the meals planned for day 3 provide 1–3 cups of vegetables, 1–2 cups of fruit, 3–8 oz of grains with half from whole grains, 2–6½ oz of protein foods, and 2–3 cups of dairy foods?

13. What changes, if any, need to be made to the meal plan in order to meet the recommended daily allowances?

Activity: Analyzing Nutrition Facts Labels

Choosing a nutritious snack food can be difficult when there are so many high-calorie choices available. Many of these are also high in fat and sodium. Convenience often plays a factor in snack choices. Nutrition facts labels found on all packaged foods and beverages list the nutrition value of the product and provide a standard for comparing foods based on nutrition.

Compare nutrition facts labels and then complete the following sentences.

____________________ **1.** Butter flavored microwave popcorn has ___ g of fat per serving.

____________________ **2.** Butter flavored microwave popcorn has ___ calories per serving.

____________________ **3.** A healthier alternative is ___.

____________________ **4.** Alternative snack has ___ g of fat per serving.

____________________ **5.** Alternative snack has ___ calories per serving.

____________________ **6.** Potato chips have ___ g of fat per serving.

____________________ **7.** Potato chips have ___ calories per serving.

____________________ **8.** A healthier alternative is ___.

____________________ **9.** Alternative snack has ___ g of fat per serving.

____________________ **10.** Alternative snack has ___ calories per serving.

____________________ **11.** Corn chips have ___ g of fat per serving.

____________________ **12.** Corn chips have ___ calories per serving.

____________________ **13.** A healthier alternative is ___.

____________________ **14.** Alternative snack has ___ g of fat per serving.

____________________ **15.** Alternative snack has ___ calories per serving.

____________________ **16.** Cheese crackers have ___ g of fat per serving.

____________________ **17.** Cheese crackers have ___ calories per serving.

____________________ **18.** A healthier alternative is ___.

____________________ **19.** Alternative snack has ___ g of fat per serving.

____________________ **20.** Alternative snack has ___ calories per serving.

____________________ **21.** Chocolate sandwich cookies with creme filling have ___ g of fat per serving.

____________________ **22.** Chocolate sandwich cookies with creme filling have ___ calories per serving.

____________________ **23.** A healthier alternative is ___.

__________ **24.** Alternative snack has ___ g of fat per serving.

__________ **25.** Alternative snack has ___ calories per serving.

__________ **26.** Commercially prepared chocolate chip cookies have ___ g of fat per serving.

__________ **27.** Commercially prepared chocolate chip cookies have ___ calories per serving.

__________ **28.** A healthier alternative is ___.

__________ **29.** Alternative snack has ___ g of fat per serving.

__________ **30.** Alternative snack has ___ calories per serving.

__________ **31.** A chewy chocolate chip granola bar has ___ g of fat per serving.

__________ **32.** A chewy chocolate chip granola bar has ___ calories per serving.

__________ **33.** A healthier alternative is ___.

__________ **34.** Alternative snack has ___ g of fat per serving.

__________ **35.** Alternative snack has ___ calories per serving.

__________ **36.** Commercially prepared powdered sugar donuts have ___ g of fat per serving.

__________ **37.** Commercially prepared powdered sugar donuts have ___ calories per serving.

__________ **38.** A healthier alternative is ___.

__________ **39.** Alternative snack has ___ g of fat per serving.

__________ **40.** Alternative snack has ___ calories per serving.

__________ **41.** A 12 oz can of regular soda has ___ g of fat per serving.

__________ **42.** A 12 oz can of regular soda has ___ calories per serving.

__________ **43.** A healthier alternative is ___.

__________ **44.** Alternative beverage has ___ g of fat per serving.

__________ **45.** Alternative beverage has ___ calories per serving.

46. Prepare and present a report about nutrition fact label findings to the class. Include information on other possible snack choices. Use visuals, such as pictures, spreadsheets, manufacturer brochures, etc., to enhance the report.

Activity: Matching Nutrient Functions

Nutrition is the study of how food is taken in and utilized by the human body. A nutrient is a substance found in food that is necessary for the body to function properly. Nutrients include proteins, carbohydrates, lipids, vitamins, minerals, and water. Each nutrient performs essential functions.

Match each nutrient to its essential function.

____________________ **1.** Proteins

____________________ **2.** Carbohydrates

____________________ **3.** Lipids

____________________ **4.** Vitamins and minerals

____________________ **5.** Water

A. Help to regulate body processes such as cell growth, reproduction, immunity, and fluid balance

B. Help the body absorb nutrients, provide insulation, manufacture hormones, and cushion organs; used as a second resource for energy

C. Build and repair body tissue

D. Transports nutrients, carries away waste, provides moisture, and helps normalize body temperature

E. Provide the body's main source of energy

Activity: Researching Vitamins or Minerals

Each vitamin and mineral plays a vital role in promoting health and well-being. Eating a wide variety of foods everyday ensures the body receives a sufficient amount of vitamins and minerals.

Use the following outline to research a vitamin or mineral.

I. History
   A. Date of discovery
   B. Individual who discovered the nutrient
II. Function
III. Food sources
IV. Recommended daily intake
V. Symptoms of deficiency
VI. Symptoms of excess
VII. Recent scientific development

1. Prepare a report and present the findings to the class using the outline above. Use visuals, such as pictures, spreadsheets, charts, graphs, and food demonstrations, to enhance the report. Include a list of sources used for the project.

Activity: Analyzing Menu Selections

Many restaurants offer menu substitutions in the interest of health. Some foodservice operations may have dietary information available to help guests identify nutritional content, preparation method, and/or ingredients used.

Complete the following activity using a menu from a local foodservice operation or an online resource.

1. Select an appetizer. Read the menu description and identify the ingredients and method of preparation.

2. Identify how the ingredients and/or preparation could be altered in order to make the appetizer healthier.

3. How can changing the ingredients and/or preparation for the appetizer make it a healthier option?

4. Select a salad and salad dressing. Read the menu description and identify the ingredients and method of preparation.

5. Identify how the ingredients and/or preparation could be altered in order to make the salad and salad dressing healthier.

6. How can changing the ingredients and/or preparation for the salad and salad dressing make it a healthier option?

7. Select an entrée. Read the menu description and identify the ingredients and method of preparation.

8. Identify how the ingredients and/or preparation could be altered in order to make the entrée healthier.

9. How can changing the ingredients and/or preparation for the entrée make it a healthier option?

10. Select a side dish. Read the menu description and identify the ingredients and method of preparation.

11. Identify how the ingredients and/or preparation could be altered in order to make the side dish healthier.

12. How can changing the ingredients and/or preparation for the side dish make it a healthier option?

13. Select a dessert. Read the menu description and identify the ingredients and method of preparation.

14. Identify how the ingredients and/or preparation could be altered in order to make the dessert healthier.

15. How can changing the ingredients and/or preparation for the dessert make it a healthier option?

Activity: Revising Menus

Consumers often seek menus offering meals that are both nutritious and full of flavor. To meet consumer demand for healthy meals, menus may require revisions.

Complete the following:

1. Create a revised menu containing healthier choices based on the information from Activity: Analyzing Menu Selections. Explain why the changes were made and the impact these changes have on nutritional content.

Name ______________________________ Date ______________

True-False

T F **1.** Reduction is the process of gently simmering a liquid until it lessens in volume and results in a thicker liquid with a less concentrated flavor.

T F **2.** Convection heat transfer is the reason that fat is a consistent temperature throughout a fryer once it reaches a set temperature.

T F **3.** Metal is a good conductor of heat, while other materials, such as glass, wood, and plastic, are not.

T F **4.** An impinger conveyor oven operates on radiation heat transfer.

T F **5.** Searing is the process of slowly cooking food to soften its texture.

Multiple Choice

______________ **1.** The Maillard reaction is a reaction that occurs when the ___ and sugars in a food are exposed to heat and merge together to form a brown exterior surface.

A. proteins
B. starches
C. lipids
D. nutrients

______________ **2.** ___ radiation can be seen in a toaster or a broiler when the heating element glows red.

A. Infrared
B. Induction
C. Convection
D. Microwave

______________ **3.** ___ is a type of heat transfer that occurs due to the circular movement of a fluid or a gas.

A. Electromagnetism
B. Radiation
C. Convection
D. Conduction

______________ **4.** ___ is the process of slowly cooking food to soften its texture.
A. Coagulation
B. Reduction
C. Searing
D. Sweating

______________ **5.** ___ is the process of a heated starch absorbing a liquid and swelling, which thickens the liquid.
A. Coagulation
B. Reduction
C. Caramelization
D. Gelatinization

Completion

______________ **1.** ___ is energy that is transferred between two objects or substances of different temperatures.

______________ **2.** Heating a pot of water is an example of ___ heat transfer.

______________ **3.** Microwave radiation uses ___ waves to heat the water, fat, and sugar molecules in food.

______________ **4.** ___ is the process of a protein changing from a liquid to a semisolid or a solid state when heat or friction is applied.

______________ **5.** Adding ___ to a stock thickens the stock and forms a sauce as it absorbs some of the water from the stock.

______________ **6.** Caramelization is a reaction that occurs when ___ are exposed to high heat and produce browning and a change in flavor.

SECTION 7.2 DRY-HEAT COOKING

Name ______________________________ **Date** ________________

True-False

T F **1.** To broil food, tongs are used to place the items in the broiler with the presentation side facing up.

T F **2.** Baking is the primary cooking method used to cook meat, poultry, and seafood.

T F **3.** A French grill is a griddle with raised ridges that creates grill marks where the food touches the ridges.

T F **4.** Pan-frying is the process of quickly cooking items in a heated wok with a very small amount of fat while constantly stirring the items.

Multiple Choice

______________ **1.** Breading is a three-step procedure used to coat and seal an item before it is ___.
- A. barbequed
- B. stewed
- C. steamed
- D. fried

______________ **2.** ___ is a dry-heat cooking method in which food is cooked on open grates above a direct heat source.
- A. Smoking
- B. Grilling
- C. Frying
- D. Griddling

______________ **3.** ___ is the process of lightly dusting an item in seasoned flour or fine bread crumbs.
- A. Dredging
- B. Stir-frying
- C. Battering
- D. Griddling

Completion

______________ **1.** ___ is a dry-heat cooking method in which food is cooked directly under a heat source.

______________ **2.** The higher the ___, the better suited the oil is for frying.

______________ **3.** Frying is a(n) ___ cooking method in which food is cooked in hot fat over moderate to high heat.

______________ **4.** In the ___ method, an item is slowly lowered into hot fat without the use of a fryer basket.

______________ **5.** ___ is a similar cooking process to smoking, but it is done on a range top.

______________ **6.** Carryover cooking helps a cooked item retain ___ that will escape as steam if the item is cut too soon after cooking.

Matching

______________ **1.** Frying

______________ **2.** Broiling

______________ **3.** Baking

______________ **4.** Sautéing

______________ **5.** Roasting

______________ **6.** Grilling

A. Dry-heat cooking method in which food is cooked uncovered in an oven

B. Dry-heat cooking method in which food is cooked directly under a heat source

C. Dry-heat cooking method in which food is cooked in hot fat over moderate to high heat

D. Dry-heat cooking method in which food is cooked uncovered in an oven or on a revolving spit over an open flame

E. Dry-heat cooking method in which food is cooked on open grates above a direct heat source

F. Dry-heat cooking method in which food is cooked quickly in a sauté pan over direct heat using a small amount of fat

SECTION 7.3 MOIST-HEAT COOKING

Name ______________________________ **Date** ____________________

True-False

T F **1.** Blanching is a moist-heat cooking method in which food is briefly parcooked and then shocked by placing it in ice-cold water to stop the cooking process.

T F **2.** A mirepoix is typically comprised of 50% onions and 50% carrots that have been roughly chopped.

T F **3.** Food is poached in liquid that is held between 185°F and 205°F.

T F **4.** Eggs are commonly poached in lightly salted water and vinegar.

Multiple Choice

______________ **1.** Simmering occurs when the temperature of the cooking liquid is between ___.
A. 160°F and 180°F
B. 185°F and 205°F
C. 210°F and 230°F
D. 300°F and 350°F

______________ **2.** Potatoes and pasta are often ___ in the professional kitchen.
A. broiled
B. blanched
C. boiled
D. fried

______________ **3.** ___ is a technique in which food is steamed in a parchment paper package as it bakes in an oven.
A. Sous vide
B. Poêléing
C. Parcooking
D. En papillote

______________ **4.** Depending on the liquid, the boiling point may be more or less than ___°F (the boiling point of water at sea level).
A. 212
B. 225
C. 250
D. 265

Completion

______________________ **1.** ___ poaching requires food to be completely covered by the poaching liquid.

______________________ **2.** A(n) ___ is an ingredient added to a food to enhance its natural flavors and aromas.

______________________ **3.** Foods, such as asparagus and broccoli, are often ___ to intensify their green color.

______________________ **4.** ___ cooking is any cooking method that uses liquid or steam as the cooking medium.

Matching

______________________ **1.** Blanching

______________________ **2.** Poaching

______________________ **3.** Steaming

A. Moist-heat cooking method in which food is placed in a container that prevents steam from escaping

B. Moist-heat cooking method in which food is briefly parcooked and then shocked by placing it in ice-cold water to stop the cooking process

C. Moist-heat cooking method in which food is cooked in a liquid that is held between 160°F and 180°F

Name ______________________________ Date ________________

True-False

T F **1.** The two most common combination cooking methods are braising and stewing.

T F **2.** During braising, aromatic vegetables are added during the marinating process.

T F **3.** Sous vide causes very little shrinkage compared to traditional cooking methods.

Multiple Choice

________________ **1.** When braising, first the meat is ___ a small amount of fat.
A. battered and fried in
B. injected with
C. cooked completely in
D. seared on all sides in

________________ **2.** The equipment needed for sous vide cooking includes a ___.
A. paper envelope and an oven
B. stockpot and a deep frying pan
C. vacuum sealer and a thermal circulator
D. roasting pan with a lid

________________ **3.** Stewing requires ___.
A. more liquid than braising
B. less liquid than braising
C. large roast-size pieces of meat
D. a thermal circulator

Completion

________________ **1.** In sous vide, a thermal ___ applies additional heat as needed to maintain the cooking liquid at the desired temperature.

________________ **2.** Combination cooking methods are often used on ___ cuts of meat.

________________ **3.** ___ is a combination cooking method that is often referred to as butter roasting.

Matching

_______________ **1.** Stewing

_______________ **2.** Braising

_______________ **3.** Poêléing

_______________ **4.** Sous vide

A. Combination cooking method in which food is browned in fat and then cooked, tightly covered, in a small amount of liquid for a long time

B. Combination cooking method in which bite-sized pieces of food are barely covered with a liquid and simmered for a long time in a tightly covered pot

C. Combination cooking method in which poultry or meat is placed on top of aromatic vegetables in a pot and basted with butter

D. Combination cooking method for cooking vacuum-sealed food by maintaining a low temperature and warming food gradually to a set temperature

Name ______________________________ **Date** ________________

True-False

T F **1.** Salts, peppercorns, and citrus zest are considered flavorings because they intensify or improve the flavor of food.

T F **2.** Umami is a sweet taste described as sugary and light.

T F **3.** Warm foods often have stronger smells and flavors than cold foods.

Multiple Choice

________________ **1.** A truffle is an edible ___ with a distinct aroma and taste.
A. fungus
B. berry
C. stem
D. seed

________________ **2.** The senses of ___ are the dominant senses used by the brain to detect flavors.
A. sight and touch
B. sight and smell
C. taste and sight
D. taste and smell

________________ **3.** If vegetables have lost color, they have also lost ___.
A. fat
B. nutrients
C. lipids
D. protein

________________ **4.** Mushrooms, tomatoes, cheeses, and seaweed have a(n) ___ taste.
A. salty
B. sweet
C. umami
D. sour

Completion

______________ **1.** ___ is the combined sensory experience of taste and smell along with visual, textural, and temperature perceptions.

______________ **2.** Taste buds are located throughout the entire mouth, but they are visible as the small bumps, known as ___, that cover the tongue.

______________ **3.** Chefs are able to build ___ by blending aromas, tastes, and textures.

Matching

______________ **1.** Flavoring

______________ **2.** Aroma

______________ **3.** Taste

______________ **4.** Texture

______________ **5.** Mouthfeel

A. A sense that is activated by receptor cells that make up the taste buds

B. A typically pleasing scent that is detected by the sense of smell

C. The appearance and feel of an item

D. Item that alters or enhances the natural flavor of food

E. The way a food or beverage feels in the mouth

Name ______________________________ **Date** ____________________

True-False

T F **1.** A chive is a leaf herb with hollow, grass-shaped, green sprouts and a mild onion flavor.

T F **2.** When substituting dried herbs for fresh herbs, a smaller amount of dried herbs should be used than the amount of fresh herbs called for in the recipe.

T F **3.** Commonly used leaf herbs include basil, chervil, cilantro, lavender, mint, rosemary, and tarragon.

T F **4.** Lavender is used in classical Italian and Greek cuisine.

Multiple Choice

______________ **1.** ___ basil is the most common type of basil and is traditionally used in tomato sauces and pesto.
A. Sour
B. Opal
C. Sweet
D. Lemon

______________ **2.** ___, also known as wild marjoram, is an herb with small, dark-green, slightly curled leaves and is a member of the mint family.
A. Parsley
B. Cilantro
C. Oregano
D. Savory

______________ **3.** ___ is an herb that has smooth, slightly elongated leaves and is best known as the flavoring in béarnaise sauce.
A. Thyme
B. Sage
C. Rosemary
D. Tarragon

______________ **4.** ___ garlic is the most common variety and the most pungent in flavor.
A. Elephant
B. Pink
C. Black
D. White

Completion

______________ **1.** If a dish has a long cooking time, fresh herbs are added at the ___ of the cooking process.

______________ **2.** ___ is one of the primary herbs used in pickling foods.

______________ **3.** Parsley has a tangy flavor that is most prominent in the ___.

______________ **4.** ___ and sage are the two strongest herbs.

Matching

______________ **1.** Herb

______________ **2.** Garlic

______________ **3.** Basil

______________ **4.** Cilantro

______________ **5.** Filé powder

______________ **6.** Mint

______________ **7.** Sage

______________ **8.** Tea

______________ **9.** Lemongrass

______________ **10.** Marjoram

A. Herb that comes from the stem and leaves of the coriander plant

B. Herb that has short, oval, pale-green leaves and is a member of the mint family

C. Bulb vegetable made up of several small cloves that are enclosed in a thin, husklike skin

D. Herb that has jagged leathery leaves and comes from an evergreen bush in the magnolia family

E. Flavoring derived from the leaves or stem of very aromatic plants

F. General term used to describe a family of similar herbs

G. Herb that has a pointy green leaf, is a member of the mint family, and is often used to add flavor to pasta

H. Herb that has narrow, velvety, green-gray leaves and is a member of the mint family

I. Stem herb with long, thin, gray-green leaves, a white scallionlike base, and a lemony flavor

J. Herb that is made from the ground leaves of the sassafras tree

Name ______________________________ **Date** ____________________

True-False

T F **1.** Examples of flower spices include allspice, cardamom, paprika, and Szechuan pepper.

T F **2.** When the roots of turmeric are ground, a bright-yellow powder is produced that is sometimes referred to as "Indian saffron."

T F **3.** A vanilla bean is the dark-brown pod of a tropical orchid.

T F **4.** Cassia is thinner and darker in color than cinnamon.

T F **5.** Root spices include cardamom, allspice, and vanilla.

Multiple Choice

______________ **1.** Although ___ seed comes from the same plant as the herb cilantro, the flavor is completely different, and the two should never be substituted for one another.
A. coriander
B. cumin
C. fennel
D. caraway

______________ **2.** ___ has a licorice flavor and is used in licorice products, sweet rolls, cookies, sweet pickles, and candies.
A. Saffron
B. Tamarind
C. Coriander
D. Anise

______________ **3.** ___ seeds are commonly used to flavor rye bread.
A. Celery
B. Achiote
C. Anise
D. Caraway

______________ **4.** ___ is the main ingredient in Worcestershire sauce.
A. Sesame seed
B. Tamarind
C. Horseradish
D. Mace

______________ **5.** Saffron is obtained from the ___ of a flower.
A. stem
B. petals
C. stigma
D. bud

______________ **6.** Coffee beans get their brown color from ___.
A. oxidation
B. the roasting process
C. nutrients in soil
D. photosynthesis

Completion

______________ **1.** Common ___ spices used in the professional kitchen include cinnamon and cassia.

______________ **2.** Mace is a spice made from the lacy red-orange covering of the ___ kernel.

______________ **3.** Because its complex flavor suggests a combination of cinnamon, nutmeg, and cloves, this pea-shaped berry spice is called ___.

______________ **4.** ___ possess the most pungent flavor of all the spices and are sometimes referred to as the nail-shaped spice because of their shape.

______________ **5.** A(n) ___ is a flavoring derived from the bark, seeds, roots, flowers, berries, or beans of aromatic plants.

SECTION 7.8 COMBINING SPICES AND HERBS

Name ______________________________ **Date** ________________

True-False

T F **1.** Cajun seasoning is used to impart the mildly sweet flavors associated with Cajun cuisine.

T F **2.** Chinese five-spice powder is typically made of equal portions of mustard powder, cloves, fennel seeds, star anise, and Szechuan pepper.

T F **3.** A variety of spices and herbs can be combined to create complementary blends that add appealing flavors to food.

Multiple Choice

______________ **1.** ___ is a mild to spicy blend that can consist of more than 20 spices including cardamom, cinnamon, cloves, coriander, cumin, fenugreek, ginger, mace, nutmeg, red and black pepper, and turmeric.
A. Chili powder
B. Curry powder
C. Cajun seasoning
D. Jerk seasoning

______________ **2.** ___ is made from ground chiles, cloves, coriander, cumin, garlic, and oregano.
A. Chili powder
B. Curry powder
C. Fines herbes
D. Jerk seasoning

______________ **3.** ___ is a spicy blend of ground allspice, chiles, cinnamon, cloves, garlic, and ginger.
A. Poultry seasoning
B. Chili powder
C. Pickling spice
D. Jerk seasoning

Completion

______________ **1.** ___ seasoning is commonly used to season blackened poultry, seafood, and meats.

______________ **2.** ___ are added to cooked sauces just before service.

______________ **3.** ___ is made of dried basil, fennel seed, lavender, marjoram, rosemary, sage, savory, and thyme and used to season roasted or grilled poultry and meats.

Name ______________________________ Date ____________________

True-False

T F **1.** Most marinades contain an acidic liquid base, such as soy sauce, wine, or vinegar.

T F **2.** Thickly coated rubs should always be wiped off the food prior to cooking.

T F **3.** Rubs, marinades, and brines allow food to absorb ingredients that can enhance the taste and texture of a finished product.

T F **4.** When a brined protein is cooked, it is especially tender because the brine aids in moisture retention.

Multiple Choice

________________ **1.** A ___ is a flavorful liquid used to soak uncooked foods such as meat, poultry, and fish to impart flavor and sometimes to tenderize.

A. brine
B. marinade
C. wet rub
D. dry rub

________________ **2.** An electric spice grinder is often used to grind and combine ___ ingredients.

A. dry rub
B. brine
C. liquid
D. acidic

________________ **3.** Marinades are commonly mixed in a deep stainless steel pan or a ___.

A. stockpot
B. hotel pan
C. plastic storage bag
D. double boiler

________________ **4.** A ___ is also known as a paste.

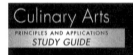

A. marinade
B. brine
C. dry rub
D. wet rub

Completion

____________________ **1.** A brine is a salt solution that usually consists of ___ of salt per 1 gal. of water.

____________________ **2.** Once a raw protein, such as poultry, has been removed from a marinade, the marinade should be ___ to prevent cross-contamination.

____________________ **3.** ___ are used to preserve foods such as herring and pickles.

____________________ **4.** Marinades and brines are always ___ based.

Name ______________________________ **Date** ________________

True-False

T F **1.** Ketchup is commonly used as a condiment for sandwiches and French fries.

T F **2.** Because prepared yellow mustard has a high acid content, it does not spoil.

T F **3.** Aioli is prepared like mayonnaise, but it also includes ketchup and pickle relish.

T F **4.** Mayonnaise is a thick, uncooked emulsion formed by combining oil with egg yolks and vinegar or lemon juice.

Multiple Choice

________________ **1.** ___ mustard is a type of yellow mustard that has a hot, spicy flavor and is made from ground yellow and brown or black mustard seeds, wheat flour, and turmeric.

A. Prepared
B. Dijon
C. English
D. American

________________ **2.** Because ___ are two of the main ingredients in ketchup, it has a slightly sweet-and-sour acidic flavor.

A. garlic and sugar
B. salt and turmeric
C. lemon and oil
D. vinegar and sugar

________________ **3.** Aioli is a popular condiment that is similar to ___.

A. mayonnaise
B. Dijon mustard
C. ketchup
D. English mustard

________________ **4.** ___ is a pungent powder or paste made from a plant with the same name.

A. Ketchup
B. Mustard
C. Mayonnaise
D. Hot sauce

Completion

_______________ **1.** A(n) ___ is a savory, sweet, spicy, or salty accompaniment that is added to or served with a food to impart a particular flavor that will complement the dish.

_______________ **2.** ___ gives prepared yellow mustard its color.

_______________ **3.** ___ mustard is a light-tan mustard that has a strong, tangy flavor and is made from brown or black mustard seeds, vinegar, white wine, sugar, and salt.

_______________ **4.** ___ is a thick, tomato-based product that usually includes vinegar, sugar, salt, and spices.

SECTION 7.11 CONDIMENT SAUCES

Name ______________________________ **Date** ________________

True-False

T F **1.** Soy sauce is a type of Asian sauce made from mashed soybeans, wheat, salt, and water.

T F **2.** A marinade is an accompaniment that is served with a food to complete or enhance the flavor and moistness.

T F **3.** The burn that is felt when pepper sauces are consumed is caused by capsaicin.

T F **4.** Soy sauce is darker, thicker, and milder in taste than tamari.

T F **5.** The most common sauce featuring chipotle peppers is called adobo sauce.

Multiple Choice

______________ **1.** ___ sauce is a type of pepper sauce made from dried, smoked jalapeño peppers.
A. Tabasco®
B. Sriracha
C. Chipotle
D. Cayenne

______________ **2.** ___ sauce is a type of Asian sauce made from fermented soybean paste, garlic, vinegar, chiles, and sugar.
A. Hoisin
B. Oyster
C. Soy
D. Asian chili

______________ **3.** When soy sauce is produced, it is fermented in vats for up to ___ months and is then pressed and strained.
A. 6
B. 12
C. 18
D. 24

_______________ **4.** ___ sauce is a type of sauce traditionally made with anchovies, garlic, onions, lime, molasses, tamarind, and vinegar.
A. Hoisin
B. Tabasco®
C. Oyster
D. Worcestershire

_______________ **5.** ___ sauce, also known as Louisiana-style hot sauce, is a type of pepper sauce made from cayenne peppers, vinegar, and salt.
A. Sriracha
B. Cayenne
C. Tabasco®
D. Chipotle

_______________ **6.** ___ are used to measure the heat level of hot peppers.
A. Scoville units
B. Calories
C. Drops
D. Teaspoons

_______________ **7.** Fish sauce has a pungent, ___ flavor that adds depth of flavor to dishes.
A. bitter
B. earthy
C. sour
D. salty

_______________ **8.** Fermented black bean sauce is commonly added to ___ dishes near the end of the cooking process.
A. Asian
B. French
C. Greek
D. English

Completion

_______________ **1.** A(n) ___ sauce is a type of sauce used to baste cooked protein and is often made with tomatoes, onions, mustard, garlic, brown sugar, and vinegar.

_______________ **2.** ___ is found in the seeds and the veins or membranes of a chili pepper.

_______________ **3.** Asian chili sauce has a spicy, savory flavor and is marketed as ___, sambal, or sambal oelek depending on where it is produced.

_______________ **4.** A(n) ___ is a cooked, mayonnaise-like product usually made from distilled vinegar, vegetable oil, water, sugar, mustard, salt, modified corn flour, and emulsifiers.

_______________ **5.** A(n) ___ is a type of hot and spicy sauce made from various types of hot peppers and typically includes vinegar.

Name ______________________________ **Date** ________________

True-False

T F **1.** Chestnuts are higher in starch and lower in fat than any other nut.

T F **2.** Some nuts, such as peanuts and soy nuts, are not nuts.

T F **3.** A chestnut is also known as a filbert.

Multiple Choice

______________ **1.** ___ are never sold in their shells due to toxins that may be present.
A. Brazil nuts
B. Hazelnuts
C. Almonds
D. Cashews

______________ **2.** A ___ is a legume that is contained in a thin, netted, tan-colored pod that grows underground.
A. peanut
B. pecan
C. soy nut
D. pistachio

______________ **3.** The ___ walnut has a strong, somewhat bitter flavor.
A. ebony
B. English
C. black
D. Brazilian

Completion

______________ **1.** A(n) ___ is a hard-shelled, dry fruit or seed that contains an inner kernel.

______________ **2.** Macadamia nuts have a very ___ fat content and a sweet, creamy taste.

______________ **3.** A(n) ___ is the edible seed of a nonwoody plant and grows in multiples within a pod.

Name ______________________________ **Date** ____________________

True-False

T F **1.** Peppercorns vary in color as a direct result of when they are harvested and how they are processed.

T F **2.** Olive oil is best used to season foods or for deep-frying.

T F **3.** Kosher salt crystals are larger than other salts and contain high concentrations of iodine.

T F **4.** Canola oil has a neutral flavor and a high smoke point, making it ideal for frying foods.

Multiple Choice

______________ **1.** A ___ salt is salt produced through the evaporation of seawater.
A. kosher
B. sea
C. pickling
D. curing

______________ **2.** ___ is a white-colored sea salt from the Normandy coast in France.
A. Flor blanca sea salt
B. Fleur de sel
C. Cyprus flake sea salt
D. Trapani sea salt

______________ **3.** ___ salt is a pure form of salt that contains no residual dust, iodine, or other additives.
A. Pickling
B. Curing
C. Kosher
D. Sea

______________ **4.** ___ peppercorns are immature berries that are picked well before that have ripened.
A. Green
B. Black
C. Pink
D. Blue

______________ **5.** ___ vinegar is often served with fish and chips.
A. Champagne
B. Balsamic
C. Rice
D. Malt

______________ **6.** ___ vinegar is most often used in pickling and preserving foods due to its high acidic quality.
A. Distilled
B. Wine
C. Malt
D. Sherry

______________ **7.** ___ peppercorns are not actual peppercorns. They are the dried berries from the baies rose plant.
A. Red
B. Scarlet
C. Rose
D. Pink

______________ **8.** ___ oil has a unique cooking property in that it will not absorb the flavors of other foods.
A. Corn
B. Peanut
C. Canola
D. Olive

Completion

______________ **1.** ___ has the unique ability to make sweet items taste more full-bodied, sour items taste more pronounced, and bland items taste much more flavorful.

______________ **2.** ___ is the colored, outermost layer of the peel of a citrus fruit and contains a high concentration of oil.

______________ **3.** ___ is a vinegar made by aging red wine vinegar in wooden vats for many years.

Matching

______________ **1.** Seasoning

______________ **2.** Peppercorn

______________ **3.** Vinegar

______________ **4.** Oil

A. Sour, acidic liquid made from fermented alcohol

B. Dried berry of a climbing vine known as the Piper nigrum and is used whole, ground, or crushed

C. Type of fat that remains in a liquid state at room temperature

D. Ingredient used to intensify or improve the natural flavor of foods

CHAPTER 7 ACTIVITIES
COOKING TECHNIQUES

Name ______________________________ **Date** ______________

Activity: Determining Types of Heat Transfer

Cooking methods rely on dry heat, moist heat, and combination heat. Depending on the method used, heat transfer occurs through conduction, convection, or radiation. Conduction uses direct physical contact to transfer heat. Convection uses a medium surrounding the food such as hot air, hot water, or hot fat to transfer heat. Radiation transfers heat through heat waves.

Identify the type of heat used (dry, moist, or combination) and explain how heat is transferred (conduction, convection, or radiation) for the following cooking methods.

______________ **1.** Poaching uses ___ heat.

______________ **2.** When poaching, heat is transferred through ___.

______________ **3.** Grilling uses ___ heat.

______________ **4.** When grilling, heat is transferred through ___.

______________ **5.** Stir-frying uses ___ heat.

______________ **6.** When stir-frying, heat is transferred through ___.

______________ **7.** Roasting uses ___ heat.

______________ **8.** When roasting, heat is transferred through ___.

______________ **9.** Simmering uses ___ heat.

______________ **10.** When simmering, heat is transferred through ___.

______________ **11.** Deep-frying uses ___ heat.

______________ **12.** When deep-frying, heat is transferred through ___.

______________ **13.** Stewing uses ___ heat.

______________ **14.** When stewing, heat is transferred through ___.

______________ **15.** Sautéing uses ___ heat.

______________ **16.** When sautéing, heat is transferred through ___.

______________ **17.** Broiling uses ___ heat.

______________ **18.** When broiling, heat is transferred through ___.

______________ **19.** Pan-frying uses ___ heat.

______________ **20.** When pan-frying, heat is transferred through ___.

______________ **21.** Boiling uses ___ heat.

______________ **22.** When boiling, heat is transferred through ___.

______________ **23.** Steaming uses ___ heat.

______________ **24.** When steaming, heat is transferred through ___.

______________ **25.** Braising uses ___ heat.

______________ **26.** When braising, heat is transferred through ___.

______________ **27.** Blanching uses ___ heat.

______________ **28.** When blanching, heat is transferred through ___.

______________ **29.** Baking uses ___ heat.

______________ **30.** When baking, heat is transferred through ___.

Activity: Analyzing Cooking Effects

When heat is introduced to food through the cooking process, characteristics of the food begin to change. Proteins, carbohydrates, fats, vitamins, minerals, and water content are altered when exposed to heat and this can cause a change in the weight, color, texture, or flavor of the food.

Obtain two chicken breasts and weigh each one.

1. How much does each chicken breast weigh?

Roast both chicken breasts in a 325°F oven until an internal temperature of 165°F is met. Slice one chicken breast immediately into ½-inch slices. Weigh the slices from the sliced chicken breast.

2. How much does the sliced chicken breast weigh after cooking?

Allow the second chicken breast to rest for 5 minutes. Weigh the whole chicken breast after the rest period.

3. How much does the unsliced chicken breast weigh after cooking?

4. Why is there a difference in weight between the two chicken breasts?

Taste each of the chicken breasts.

5. Describe the texture and flavor of each chicken breast.

6. If the texture and flavor of the chicken breasts are different, what is the cause?

Shape two hamburger patties from ground beef. Weigh each hamburger patty to make sure they are equal in weight.

7. How much does each hamburger patty weigh?

Cook one hamburger patty on high heat until an internal temperature of 160°F is met. Press down on the meat with a spatula to flatten the hamburger patty while it cooks.

8. How much does the pressed hamburger patty weigh after cooking?

Cook the second hamburger patty on medium heat until an internal temperature of 160°F is met. Do not press the second hamburger patty as it cooks.

9. How much does the hamburger patty that was not pressed weigh after cooking?

10. Why is there a difference in weight between the two hamburger patties?

Taste each of the hamburger patties.

11. Describe the texture and flavor of each hamburger patty.

12. If the texture and flavor of the hamburger patties are different, what is the cause?

Slice an onion. Heat a sauté pan and add 1 tbsp oil to the pan. Sweat the onions in the hot pan. Taste the sweated onions.

13. Describe the flavor of the sweated onions.

Sauté the onions to a light golden brown. Taste the lightly sautéed onions.

14. Describe the flavor of the lightly sautéed onions.

Further sauté the onions until they turn a rich caramel brown color. Taste the caramelized onions.

15. Describe the flavor of the caramelized onions.

16. Why is there a difference between the flavor of the lightly sautéed onions and the caramelized onions?

Obtain two pork chops. Place one pork chop in a sauté pan with a small amount of water, cover the pan, and simmer the pork chop until an internal temperature of 145°F is met and held for 3 minutes. Taste the simmered pork chop.

17. Describe the appearance and flavor of the simmered pork chop.

Sear both sides of the second pork chop using a dry-heat cooking method. Add a small amount of water to the pan, cover the pan, and simmer the pork chop until an internal temperature of 145°F is met and held for 3 minutes. Taste the seared and simmered pork chop.

18. Describe the appearance and flavor of the seared and simmered pork chop.

19. If the appearance and flavor of the pork chops are different, what is the cause?

Obtain two small chuck roasts. Season and braise one chuck roast until fork tender. Taste the braised chuck roast.

20. Describe the texture and flavor of the braised chuck roast.

Season and grill the second chuck roast until an internal temperature of 145°F is met and held for 3 minutes. Taste the grilled chuck roast.

21. Describe the texture and flavor of the grilled chuck roast.

22. Why is there a difference in the texture and flavor of the two roasts?

Activity: Examining Sensory Perception

The senses of taste and smell are closely linked. The smell or aroma of a food can often influence what a person tastes. The four major types of flavors that the tongue can clearly identify are salty, sweet, bitter, and sour.

Blindfold a lab partner and ask the partner to hold his or her nose. Place a drop of lime juice on the lab partner's tongue with a straw.

1. How does the lab partner describe the taste?

Have the blindfolded lab partner drink water. The blindfolded lab partner should not hold his or her nose. Place a drop of lime juice on the lab partner's tongue with a straw.

2. How does the lab partner describe the taste?

3. Can the lab partner taste more flavor when holding or when not holding his or her nose?

Have the blindfolded lab partner drink water. Have the blindfolded lab partner hold his or her nose. Place a drop of lemon juice on the lab partner's tongue with a straw.

4. How does the lab partner describe the taste?

Have the lab partner drink water. The blindfolded lab partner should not hold his or her nose. Place a drop of lemon juice on the lab partner's tongue with a straw.

5. How does the lab partner describe the taste?

6. Can the lab partner taste more flavor when holding or when not holding his or her nose?

Have the blindfolded lab partner drink water. Have the blindfolded lab partner hold his or her nose. Place a drop of orange juice on the lab partner's tongue with a straw.

7. How does the lab partner describe the taste?

Have the lab partner drink water. The blindfolded lab partner should not hold his or her nose. Place a drop of orange juice on the lab partner's tongue with a straw.

8. How does the lab partner describe the taste?

9. Can the lab partner taste more flavor when holding or when not holding his or her nose?

Have the blindfolded lab partner drink water. Have the blindfolded lab partner hold his or her nose. Place a drop of apple juice on the lab partner's tongue with a straw.

10. How does the lab partner describe the taste?

Have the lab partner drink water. The blindfolded lab partner should not hold his or her nose. Place a drop of apple juice on the lab partner's tongue with a straw.

11. How does the lab partner describe the taste?

12. Can the lab partner taste more flavor when holding or when not holding his or her nose?

Have the lab partner drink water. The blindfolded lab partner should not hold his or her nose. Place a small amount of salt on the blindfolded lab partner's tongue with a coffee stirrer.

13. How does the lab partner describe the taste?

14. What area of the tongue can taste the salt?

Have the lab partner drink water. The blindfolded lab partner should not hold his or her nose. Place a small amount of lemon juice on the blindfolded lab partner's tongue with a coffee stirrer.

15. How does the lab partner describe the taste?

16. What area of the tongue can taste the lemon juice?

Have the lab partner drink water. The blindfolded lab partner should not hold his or her nose. Place a small amount of cocoa powder on the blindfolded lab partner's tongue with a coffee stirrer.

17. How does the lab partner describe the taste?

18. What area of the tongue can taste the cocoa powder?

Have the lab partner drink water. The blindfolded lab partner should not hold his or her nose. Place a small amount of sugar on the blindfolded lab partner's tongue with a coffee stirrer.

19. How does the lab partner describe the taste?

20. What area of the tongue can taste the sugar?

Activity: Identifying Herbs and Spices

A chef must be able to identify herbs and spices by appearance and smell. The appearance and smell of herbs and spices are also important to determine freshness and quality of ingredients.

Match each herb or spice with its image.

__________ **1.** Cinnamon

__________ **2.** Fennel seed

__________ **3.** Bay leaves

__________ **4.** Vanilla bean

__________ **5.** Nutmeg

__________ **6.** Chives

__________ **7.** Cilantro

__________ **8.** Ginger

__________ **9.** Saffron

__________ **10.** Dill

__________ **11.** Cloves

__________ **12.** Sage

__________ **13.** Rosemary

__________ **14.** Lemon grass

__________ **15.** Basil

Ⓐ

Shenandoah Growers
Ⓑ

Shenandoah Growers
Ⓒ

Ⓓ

Ⓔ

Shenandoah Growers
Ⓕ

Ⓖ

Ⓗ

Shenandoah Growers
Ⓘ

Ⓙ

Shenandoah Growers
Ⓚ

Shenandoah Growers
Ⓛ

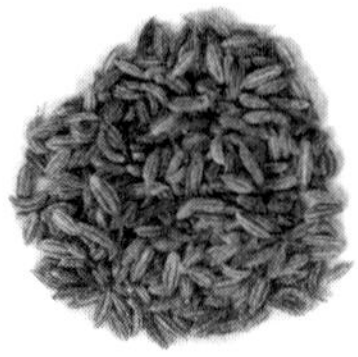

Ⓜ

Shenandoah Growers
Ⓝ

Shenandoah Growers
Ⓞ

Activity: Matching Herb and Spice Terms

Herbs and spices can be used in creative combinations to develop inspiring dishes. It is important to be able to use herbs that blend well and enhance flavors.

Match each herb with its description.

______ **1.** Basil

______ **2.** Bay leaves

______ **3.** Chives

______ **4.** Cilantro

______ **5.** Dill

______ **6.** Mint

______ **7.** Parsley

______ **8.** Rosemary

______ **9.** Sage

______ **10.** Tarragon

______ **11.** Oregano

______ **12.** Thyme

______ **13.** Filé powder

______ **14.** Lemongrass

A. Thick, aromatic leaf from the evergreen bay laurel tree

B. Stem herb with hollow, grass-shaped, green sprouts and a mild onion flavor

C. Herb that has feathery, blue-green leaves and is a member of the parsley family

D. Herb that has needlelike leaves and is a member of the evergreen family

E. Herb that has curly or flat dark-green leaves and is used as both a flavoring and a garnish

F. Herb that has a pointy green leaf and is a member of the mint family; traditionally used in tomato sauces and pesto

G. A general term used to describe a family of similar herbs

H. Herb that has narrow, velvety, green-gray leaves and is a member of the mint family

I. Herb that comes from the stem and leaves of the coriander plant

J. Herb that is made from the ground leaves of the sassafras tree

K. Herb that has small, dark green, slightly curled leaves, is a member of the mint family, and has a slightly stronger flavor than marjoram

L. Stem herb with long, thin, gray-green leaves, a white scallionlike base, and a lemony flavor

M. Herb that has very small gray-green leaves, is a member of the mint family, and is a key ingredient in a classic bouquet garni

N. Herb that has smooth, slightly elongated leaves and is best known as the flavoring in béarnaise sauce

Match each spice with its description.

______ **15.** Allspice

______ **16.** Caraway seeds

______ **17.** Cayenne pepper

______ **18.** Cinnamon

______ **19.** Cloves

______ **20.** Curry powder

______ **21.** Ginger

______ **22.** Horseradish

______ **23.** Nutmeg

______ **24.** Saffron

______ **25.** Sesame seeds

______ **26.** Vanilla bean

O. Spice made from dried, ground berries of certain varieties of hot peppers; also known as red pepper

P. Spice made from the dried, unopened bud of a tropical evergreen tree

Q. Small, flat, white or black seed often baked on rolls, bread, and buns

R. Small, crescent-shaped brown seed that is used as a spice

S. Spice made from the dried, unripe fruit of a small pimiento tree with a flavor suggesting a combination of cinnamon, nutmeg, and cloves

T. Spice made from the bumpy root of a tropical plant grown in China and is used fresh, dried, ground, crystallized, candied, and pickled

U. Dark-brown pod of a tropical orchid commonly available fresh, dried and ground, or as a liquid extract

V. Spice from the large brown-skinned root of a shrub related to the radish

W. Spice made from the dried, bright-red stigmas of the purple crocus flower

X. Mild to spicy blend that can consist of more than 20 spices

Y. A spice made from an oval, gray-brown seed found in the yellow, nectarine-shaped fruit of a tropical evergreen; the sister spice of mace

Z. Spice made from the dried, thin, inner bark of a small evergreen tree

Activity: Researching Herbs and Spices

Herbs and spices have a long and varied history. They have been used throughout the ages for both their flavor and medicinal qualities.

Research an herb or spice and use the following outline to organize the research information.

I. History
II. Medical uses
III. Culinary uses
IV. Current trends

1. Prepare a report and present findings to the class using the outline. Use visuals, such as pictures and food demonstrations, to enhance the report. Include a list of sources used for the project.

Activity: Matching Condiments and Condiment Sauces

Condiments and condiment sauces range in flavor from salty and savory to sweet and spicy. They are most often served as an accompaniment to a food to impart or enhance flavor and/or to add moisture to a dish.

Match each of the following condiments and condiment sauces to the correct definition.

______ **1.** Oyster sauce

______ **2.** Salad dressing

______ **3.** Barbeque sauce

______ **4.** Fermented black bean sauce

______ **5.** Worcestershire sauce

______ **6.** Soy sauce

______ **7.** Ketchup

______ **8.** Dijon mustard

______ **9.** Mayonnaise

______ **10.** Fish sauce

______ **11.** English mustard

______ **12.** Pepper sauce

______ **13.** Asian chili sauce

______ **14.** Prepared yellow mustard

______ **15.** Hoisin sauce

A. Type of hot and spicy sauce made from various types of hot peppers and typically includes vinegar

B. Light-tan mustard that has a strong, tangy flavor and is made from brown or black mustard seeds, vinegar, white wine, sugar, and salt

C. Type of Asian sauce made from mashed soybeans, wheat, salt, and water

D. Type of sauce used to baste a cooked protein and is often made with tomatoes, onions, mustard, garlic, brown sugar, and vinegar

E. Red-colored Asian sauce made from puréed red chiles and garlic

F. Type of yellow mustard that has a hot, spicy flavor and is made from ground yellow and brown or black mustard seeds, wheat flour, and turmeric

G. Type of Asian sauce made from fermented and salted blackened soybeans mixed with garlic and spices

H. Type of bright-yellow mustard that is mild in flavor and made from ground yellow mustard seeds, vinegar, and turmeric

I. Type of sauce traditionally made with anchovies, garlic, onions, lime, molasses, tamarind, and vinegar

J. Type of Asian sauce made from the cooking liquid of boiled oysters, brine, and soy sauce

K. Type of Asian sauce made from fermented soybean paste, garlic, vinegar, chiles, and sugar

L. Thick, tomato-based product that usually includes vinegar, sugar, salt, and spices

M. Cooked, mayonnaise-like product usually made from distilled vinegar, vegetable oil, water, sugar, mustard, salt, modified corn flour, and emulsifiers

N. Type of Asian sauce made from a liquid drained from fermented, salted fish

O. Thick, uncooked emulsion formed by combining oil with egg yolks and vinegar or lemon juice

Activity: Identifying Nuts

Nuts can be added to many dishes to enhance both flavor and texture. Toasting nuts brings out an even richer, nuttier flavor. The versatility of nuts allows them to be used in sweet and savory applications.

Match each nut to its image.

_______________ **1.** Chestnuts

_______________ **2.** Soy nuts

_______________ **3.** Walnuts

_______________ **4.** Almonds

_______________ **5.** Pecans

_______________ **6.** Pistachios

_______________ **7.** Peanuts

_______________ **8.** Hazelnuts

_______________ **9.** Pine nuts

_______________ **10.** Macadamia nuts

_______________ **11.** Brazil nuts

_______________ **12.** Cashews

Ⓐ Ⓑ

Ⓒ Ⓓ

Trails End Chestnuts

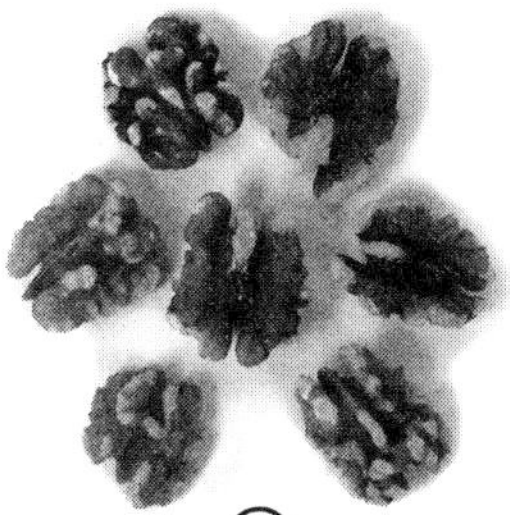

Ⓔ Ⓕ

Ⓖ Ⓗ Ⓘ

Ⓙ Ⓚ Ⓛ

Activity: Analyzing Vinegars and Oils

A seasoning is an ingredient used to intensify or improve the natural flavor of foods. When seasonings are used correctly, the flavor of the seasoning should not be noticeable. Vinegars and oils are seasonings commonly used in the professional kitchen to enhance dishes ranging from appetizers and salads to entrées and desserts.

Pour a small amount of balsamic vinegar, rice vinegar, Champagne vinegar, red wine vinegar, and cider vinegar each into five separate containers. Dip a piece of romaine lettuce into the balsamic vinegar and taste.

1. Describe the taste of the balsamic vinegar.

2. Describe how balsamic vinegar could be used in the professional kitchen.

Dip a piece of romaine lettuce into the rice vinegar and taste.

3. Describe the taste of the rice vinegar.

4. Describe how rice vinegar could be used in the professional kitchen.

Dip a piece of romaine lettuce into the Champagne vinegar and taste.

5. Describe the taste of the Champagne vinegar.

6. Describe how Champagne vinegar could be used in the professional kitchen.

Dip a piece of romaine lettuce into the red wine vinegar and taste.

7. Describe the taste of the red wine vinegar.

Dip a piece of romaine lettuce into the cider vinegar and taste.

8. Describe the taste of the cider vinegar.

9. Describe how cider vinegar could be used in the professional kitchen.

Obtain extra-virgin, virgin, pure, and olive-pomace olive oil. Pour a small amount of each olive oil onto a plate using separate plates for each. Dip a piece of romaine lettuce into the extra-virgin olive oil and taste.

10. Describe the taste of the extra-virgin olive oil.

11. Describe how extra-virgin olive oil could be used in the professional kitchen.

Dip a piece of romaine lettuce into the virgin olive oil and taste.

12. Describe the taste of the virgin olive oil.

13. Describe how virgin olive oil could be used in the professional kitchen.

Dip a piece of romaine lettuce into the pure olive oil and taste.

14. Describe the taste of the pure olive oil.

15. Describe how pure olive oil could be used in the professional kitchen.

Dip a piece of romaine lettuce into the olive-pomace oil and taste.

16. Describe the taste of the olive-pomace oil.

17. Describe how olive-pomace oil could be used in the professional kitchen.

Culinary Arts
PRINCIPLES AND APPLICATIONS
STUDY GUIDE

CHAPTER 8 REVIEW
STOCKS AND SAUCES

SECTION 8.1 STOCKS

Name ______________________ **Date** ______________

True-False

T F **1.** Bouillon is the liquid that is strained off after cooking vegetables, poultry, meat, or seafood in wine.

T F **2.** If desired in a stock, salt is added at the beginning stage of preparation.

T F **3.** Boiling a stock breaks up coagulated impurities as they are released from bones and causes the stock to become cloudy.

T F **4.** The flavor of an essence is more concentrated than a fumet and extremely aromatic.

T F **5.** The cooking time for fish stocks is much longer than white or brown stocks.

T F **6.** When properly prepared, stored, and handled, a glace will last for several months.

T F **7.** The taste of a stock should overpower the taste of the dish that it is used in.

T F **8.** The coagulated impurities of a stock should be skimmed from the surface and discarded.

Multiple Choice

______________ **1.** Once completely cooled, a(n) ___ can be cut into cubes, which are easy to store.
A. essence
B. glace
C. remouillage
D. bouillon

______________ **2.** ___ or trimmings are the primary flavoring ingredients used to make meat, poultry, or fish stocks.
A. Bones
B. Aromatics
C. Herbs
D. Spices

______________ **3.** Overcooked mirepoix can make a stock ___.
A. bitter
B. sweet
C. clear
D. cloudy

______ **4.** A fumet differs from a fish stock in that the mirepoix, bones, and shells are ___ in a stockpot prior to use.
A. boiled
B. sweated
C. roasted
D. puréed

______ **5.** A ___ is a mixture of 50% onions, 25% celery, and 25% carrots.
A. roux
B. bouquet garni
C. mirepoix
D. sachet

______ **6.** ___ water is used to start a stock.
A. Frozen
B. Cold
C. Hot
D. Boiling

Completion

______ **1.** A white matignon includes ___ instead of carrots.

______ **2.** A(n) ___ is a light-colored stock produced by gently simmering poultry, veal, or fish bones in water with vegetables and herbs.

______ **3.** A vegetable stock is a clear, light-colored stock produced by gently simmering vegetables with a(n) ___ and a sachet.

______ **4.** A(n) ___ is often used in place of water when starting a new stock as it has more flavor than plain water.

______ **5.** The most common methods for cooling stocks are in an ice bath and with a(n) ___.

Matching

______ **1.** Stock

______ **2.** Fumet

______ **3.** Matignon

______ **4.** Bouquet garni

______ **5.** Sachet d'épices

A. A bundle of herbs and vegetables tied together with twine that is used to flavor stocks and soups

B. A mixture of spices and herbs placed in a piece of cheesecloth and tied with butcher's twine

C. A concentrated stock made from fish bones or shellfish shells and vegetables

D. An unthickened liquid that is flavored by simmering seasonings with vegetables, and often, the bones of meat, poultry, or fish

E. Uniformly cut mixture of onions or leeks, carrots, and celery and may also contain smoked bacon or ham

Name ______________________________ Date ______________

True-False

T F 1. Unlike roux, a beurre manié is cooked.

T F 2. Cornstarch is used to thicken only hot foods.

T F 3. A slurry is a mixture of equal parts of hot liquid and a starch that is used to thicken other liquids.

T F 4. Reduction is the process of gently simmering a liquid until it lessens in volume and results in a thicker liquid with a more concentrated flavor.

T F 5. A roux is a thickening agent made by cooking a mixture of equal amounts, by weight, of flour and fat.

Multiple Choice

______________ 1. ___ is a thickening agent that is often combined with fat before being added to a liquid.
- A. Flour
- B. Arrowroot
- C. Roux
- D. Cornstarch

______________ 2. ___ is the consistency of a liquid that thinly coats the back of a spoon and ensures that a sauce will cling lightly to another food.
- A. Roux
- B. Foam
- C. Nappe
- D. Slurry

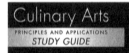

______________ 3. A ___ is added to chicken stock to create chicken gravy.
- A. thickening agent
- B. gelatin
- C. seasoning
- D. liquid

__________ **4.** A liaison is a thickening agent that is made from a mixture of ___ and heavy cream.
A. butter
B. shortening
C. egg whites
D. egg yolks

__________ **5.** A sauce thickened with arrowroot is ___ than a sauce made with cornstarch.
A. thicker
B. thinner
C. cloudier
D. clearer

Completion

___________ **1.** Gelatinization is the process of a heated ___ absorbing moisture and swelling, which thickens the liquid.

___________ **2.** A(n) ___ is an accompaniment that is served with a food to complete or enhance the flavor and moistness.

___________ **3.** ___ roux is a roux that is made by cooking equal amounts, by weight, of fat and flour until the roux exhibits a slightly golden or tan color.

___________ **4.** Egg yolks begin to coagulate at temperatures above ___°F.

___________ **5.** ___ is the white, powdery, pure starch derived from corn.

Matching

___________ **1.** Beurre manié
___________ **2.** Nappe
___________ **3.** Liaison
___________ **4.** Reduction
___________ **5.** Roux
___________ **6.** Sauce
___________ **7.** Coagulation
___________ **8.** Gelatinization

A. An accompaniment that is served with a food to complete or enhance the flavor and moistness

B. The process of a protein changing from a liquid to a semisolid or a solid state when heat or friction is applied

C. A thickening agent made by cooking a mixture of equal amounts, by weight, of flour and fat and is used to thicken sauces and soups

D. The consistency of a liquid that thinly coats the back of a spoon and ensures that a sauce will cling lightly to another food

E. The process of a heated starch absorbing moisture and swelling, which thickens the liquid

F. A thickening agent made from a mixture of egg yolks and heavy cream and is used to thicken sauces

G. A thickening agent made by kneading equal amounts, by weight, of pastry, cake, or all-purpose flour and softened butter and can be whisked into a sauce just before service

H. The process of gently simmering a liquid until it lessens in volume and results in a thicker liquid with a more concentrated flavor

Chapter 8 Review
Stocks and Sauces

Section 8.3 Classical Sauces

Name ______________________________ **Date** ________________

True-False

T F **1.** Hollandaise is a mother sauce made by thickening clarified butter with egg yolks.

T F **2.** Velouté is a mother sauce made from a flavorful white stock and a blonde roux.

T F **3.** Tomato sauce is not considered a mother sauce.

T F **4.** Egg whites contain a natural emulsifier called albumen.

T F **5.** To reduce acid in a tomato sauce, baking soda can be added while the sauce simmers.

Multiple Choice

______________ **1.** Béchamel is a mother sauce that is made by thickening milk with ___ and seasonings.
A. a white roux
B. a blonde roux
C. arrowroot
D. buerre manié

______________ **2.** A suprême sauce is a sauce made by adding cream to a ___ velouté.
A. veal
B. fish
C. chicken
D. beef

______________ **3.** A well-made béchamel should be silky smooth and have a ___ consistency.
A. watery
B. coarse
C. chunky
D. nappe

______________ **4.** A(n) ___ is a sauce made by adding fresh lemon juice and a yolk-and-cream liaison to a velouté.
A. suprême sauce
B. allemande sauce
C. espagnole
D. jus lié

______________ **5.** A disadvantage of a ___ is that it will only be as rich and flavorful as the brown stock from which it was produced because it is thickened with only a cornstarch or arrowroot slurry.
- A. jus lié
- B. demi-glace
- C. suprême sauce
- D. allemande sauce

______________ **6.** Tomato sauce immediately ___ when baking soda is added because of the chemical reactions taking place as the soda neutralizes some of the acid.
- A. burns
- B. breaks down
- C. foams
- D. clumps

Completion

______________ **1.** The five mother sauces are béchamel, velouté, espagnole, ___, and hollandaise.

______________ **2.** Espagnole is a mother sauce made from a full-bodied brown stock, ___ roux, tomato purée, and a hearty caramelized mirepoix.

______________ **3.** Typically, ___ tomatoes are used to make tomato sauce, as they have less water, fewer seeds, and more meat than other varieties.

______________ **4.** ___ is a sugar syrup made by caramelizing a small amount of granulated sugar in a saucepan and deglazing the pan with a small amount of vinegar.

______________ **5.** A(n) ___ is the most widely used middle sauce.

______________ **6.** Hollandaise relies on a(n) ___ rather than coagulation and is therefore different from other sauces thickened with yolks.

Matching

______________ **1.** Mother sauce

______________ **2.** Emulsification

______________ **3.** Onion piquet

______________ **4.** Clarified butter

______________ **5.** Jus lié

A. One of five sauces from which the small classical sauces described by Escoffier are produced; also known as a leading or grand sauce

B. Half of an onion studded with cloves and a bay leaf

C. The process of temporarily binding two liquids that do not combine easily, such as oil and vinegar

D. Butter that has been cooked long enough to allow the water within it to evaporate

E. Sauce that is made by thickening a brown stock either by adding a cornstarch or arrowroot slurry or simply by a slow reduction

Chapter 8 Review
Stocks and Sauces

Section 8.4 Butter Sauces

Name ______________________________ **Date** ____________________

True-False

T F **1.** A broken butter is a butter sauce made by freezing butter until the fat, milk solids, and water separate or "break."

T F **2.** A compound butter is a flavorful butter sauce made by mixing melted butter with flavoring ingredients.

T F **3.** Compound butters may be served alone or over grilled meats, fish, or vegetable dishes.

T F **4.** Beurre blanc sauce is slightly thicker than hollandaise and heavy cream.

Multiple Choice

______________ **1.** ___ are often prepared in advance and held in the refrigerator for a few days or in the freezer for longer periods.
- A. Foams
- B. Nages
- C. Compound butters
- D. Broken butters

______________ **2.** To prepare a(n) ___, whole butter is cooked over medium heat until the white milk solids begin to turn brown.
- A. unbroken butter
- B. compound butter
- C. beurre blanc
- D. beurre noisette

______________ **3.** ___ are prepared, rolled into a cylinder shape on a sheet of plastic wrap or parchment paper, and refrigerated until needed.
- A. Compound butters
- B. Broken butters
- C. Beurre blancs
- D. Beurre noirs

______________ **4.** A ___ is also known as a white butter sauce.
A. beurre noir
B. beurre blanc
C. broken butter
D. compound butter

Completion

______________ **1.** A(n) ___ , also known as a white butter sauce, is a butter-based emulsified sauce made by whisking cold, softened butter into a wine, white-wine vinegar, shallot, and peppercorn reduction.

______________ **2.** The key to a successful beurre blanc sauce is constantly ___ while adding the butter to the reduction.

______________ **3.** ___ is commonly referred to as "black butter."

______________ **4.** A classic dish served with a(n) ___ butter is Dover sole à la Meunière.

Chapter 8 Review
Stocks and Sauces

Section 8.5 Contemporary Sauces

Name ______________________________ **Date** ________________

True-False

T F **1.** Salsas are always puréed.

T F **2.** A neutral-flavored oil such as canola or grape seed oil is typically used to prepare a flavored oil.

T F **3.** A chutney is prepared by reserving and straining the liquid in which the main item was cooked.

T F **4.** A coulis can be served either warm or cold over grilled or sautéed items, as well as desserts.

T F **5.** Foams can be produced by making a reduction of a flavoring ingredient, shallots, garlic, and wine.

Multiple Choice

______________ **1.** A coulis is a sauce typically made from either raw or cooked puréed ___.
- A. eggs
- B. fruits
- C. herbs
- D. meats

______________ **2.** Cilantro, basil, mint, pine nuts, and Parmesan cheese are common ___ ingredients.
- A. pesto
- B. nage
- C. chutney
- D. coulis

______________ **3.** ___ are made by simply mixing diced vegetables or fruits, herbs, and spices together.
- A. Foams
- B. Nages
- C. Pestos
- D. Salsas

______ 4. The ingredients in ___ are cooked with sugar and spices to yield a sweet-and-sour flavor.
A. nages
B. pestos
C. chutneys
D. salsas

______ 5. Flavored oils can be used to add flavor, ___, color, and aroma to a dish.
A. coarseness
B. moisture
C. creaminess
D. trans fat

Completion

______ 1. A(n) ___ is an aromatic court bouillon that is often used as a finishing sauce.

______ 2. Pesto is made from fresh ingredients that have been either crushed with a mortar and pestle or finely chopped in a food processor before being mixed with ___.

______ 3. When preparing a foam, a small amount of ___ powder is added as a stabilizer and then the mixture is puréed.

______ 4. When preparing a nage, the broth is reduced up to ___ of the original volume.

______ 5. A cream whipper, or siphon canister, is a tool that uses ___ gas and pressure to create light, fluffy foams.

Name ______________________________ **Date** ________________

Activity: Analyzing Stock Basics

Stocks are used to make sauces and soups. Stocks are made primarily by simmering bones, vegetables, and seasonings in water for a length of time to extract their colors, flavors, and nutrients.

Answer the following questions.

1. What is a mirepoix?

2. What is the general composition of water, flavoring components, and mirepoix for a stock?

3. How should vegetables be prepped for a mirepoix?

4. What considerations should be made for some vegetables before adding them to a mirepoix?

5. How is a matignon different from a mirepoix?

6. What size mirepoix should be used for a chicken stock?

7. How long should a chicken stock be cooked?

8. What size of mirepoix should be used for a beef stock?

9. How long should a beef stock be cooked?

10. What size mirepoix should be used for a fish stock?

11. How long should a fish stock be cooked?

12. How long should a vegetable stock be cooked?

Activity: Thickening Stocks

There are several techniques available to thicken a stock. Each technique has an effect on the color and flavor of the completed stock. A chef must understand how each thickening method affects a stock and choose the method most appropriate for the dish being prepared.

Prepare a white roux. Add the white roux to 2 cups of chicken stock and mix well.

1. How does the white roux affect the color of the stock?

2. How does the white roux affect the flavor of the stock?

Prepare a blonde roux. Add the blonde roux to 2 cups of chicken stock and mix well.

3. How does the blonde roux affect the color of the stock?

4. How does the blonde roux affect the flavor of the stock?

Prepare a brown roux. Add the brown roux to 2 cups of chicken stock and mix well.

5. How does the brown roux affect the color of the stock?

6. How does the brown roux affect the flavor of the stock?

Prepare a buerre manié. Add the buerre manié to 2 cups of chicken stock and mix well.

7. How does the buerre manié affect the color of the stock?

8. How does the buerre manié affect the flavor of the stock?

Prepare a cornstarch slurry. Add the cornstarch slurry to 2 cups of chicken stock and mix well.

9. How does the cornstarch slurry affect the color of the stock?

10. How does the cornstarch slurry affect the flavor of the stock?

Prepare an arrowroot slurry. Add the arrowroot slurry to 2 cups of chicken stock and mix well.

11. How does the arrowroot slurry affect the color of the stock?

12. How does the arrowroot slurry affect the flavor of the stock?

Prepare a liaison. Add the liaison to 2 cups of chicken stock and mix well.

13. How does the liaison affect the color of the stock?

14. How does the liaison affect the flavor of the stock?

Answer the following questions.

15. When is a white roux used in the professional kitchen?

16. When is a blonde roux used in the professional kitchen?

17. When is a brown roux used in the professional kitchen?

18. When is a beurre manié used in the professional kitchen?

19. When is a cornstarch slurry used in the professional kitchen?

20. When is an arrowroot slurry used in the professional kitchen?

21. When is a liaison used in the professional kitchen?

Activity: Preparing Classical Sauces

Classical sauces are often referred to as mother sauces. A mother sauce is one of five sauces from which the small classical sauces described by Escoffier are produced.

Prepare the following mother sauces. Respond to the questions and statements that follow.

- Béchamel
- Velouté
- Espagnole
- Tomato
- Hollandaise

1. Describe the flavor of the prepared béchamel.

2. What color is the prepared béchamel?

3. What small sauces can be made from a béchamel?

4. Describe the flavor of the prepared velouté.

5. What color is the prepared velouté?

6. What middle sauces can be made from a velouté?

7. Describe the flavor of the prepared espagnole.

8. What color is the prepared espagnole?

9. What middle sauces can be made from an espagnole?

10. Describe the flavor of the prepared tomato sauce.

11. What color is the prepared tomato sauce?

12. What small sauces can be made from a tomato sauce?

13. Describe the flavor of the prepared hollandaise.

14. What color is the prepared hollandaise?

15. What small sauces can be made from a hollandaise?

Activity: Calculating Stock Costs

Understanding costs is an essential component for success in the foodservice industry. Knowing the cost of ingredients enables foodservice professionals to effectively track inventory, purchase items, and determine menu prices.

Cost the following recipe using the recipe costing form. Answer the questions that follow.

Recipe Costing Form								
Recipe: Beef, Veal, or Chicken Stock								
Yield: 5 gallons								
Portion Size: N/A								
Item No.	**Ingredient**	**Quantity**	**As-Purchased (AP)**		**Yield Percentage**	**Edible Portion (EP)**		**Total Cost**
			Amount	**Cost**		**Amount**	**Cost**	
488	**beef, veal, or chicken bones**	27 lb	N/A	scrap	100%	N/A	$0.00	$0.00
1	**water**	5 gal.	1 gal.	$0.01	100%			
1148	**medium onion, diced**	1 lb 13 oz	50 lb	$75.00	83%			
1120	**celery, diced**	14 oz	25 lb	$28.75	85%			
1162	**carrot, diced**	14 oz	25 lb	$28.75	92%			
1886	**salt**	1 tbsp	25 lb	$30.00	100%			
1889	**peppercorns**	1 tsp	5 lb	$40.00	100%			
1822	**bay leaf**	1 leaf	8 oz	$5.75	100%			
1832	**thyme**	1 sprig	8 oz	$6.88	100%			
1856	**parsley**	1 sprig	8 oz	$1.50	96%			
1153	**leek, julienne cut**	1 tbsp	25 lb	$35.99	96%			
							Total	

________________ **1.** The EP amount of water is ___.

________________ **2.** The EP amount of onions is ___.

________________ **3.** The EP amount of celery is ___.

________________ **4.** The EP amount of carrots is ___.

________________ **5.** The EP amount of salt is ___.

________________ **6.** The EP amount of peppercorns is ___.

________________ **7.** The EP amount of bay leaves is ___.

________________ **8.** The EP amount of thyme is ___.

________________ **9.** The EP amount of parsley is ___.

________________ **10.** The EP amount of leeks is ___.

________________ **11.** The EP cost of water is $___ per gal.

________________ **12.** The EP cost of onion is $___ per lb.

_______________ **13.** The EP cost of celery is $___ per lb.

_______________ **14.** The EP cost of carrots is $___ per lb.

_______________ **15.** The EP cost of salt is $___ per lb.

_______________ **16.** The EP cost of peppercorns is $___ per lb.

_______________ **17.** The EP cost of bay leaves is $___ per oz.

_______________ **18.** The EP cost of thyme is $___ per oz.

_______________ **19.** The EP cost of parsley is $___ per oz.

_______________ **20.** The EP cost of leeks is $___ per lb.

_______________ **21.** The total cost of water is $___.

_______________ **22.** The total cost of onions is $___.

_______________ **23.** The total cost of celery is $___.

_______________ **24.** The total cost of carrots is $___.

_______________ **25.** If 1 tbsp of salt weighs 1½ oz, the total cost of salt is $___.

_______________ **26.** If 1 tsp of peppercorn weighs ¾ oz, the total cost of peppercorns is $___.

_______________ **27.** If a bay leaf weighs ½ oz, the total cost of bay leaves is $___.

_______________ **28.** If a sprig of thyme weighs 1 oz, the total cost of thyme is $___.

_______________ **29.** If a sprig of parsley weighs ½ oz, the total cost of parsley is $___.

_______________ **30.** If 1 tbsp of leek weighs ¾ oz, the total cost of leeks is $___.

_______________ **31.** The total cost for the beef, veal, or chicken stock recipe is $___.

_______________ **32.** If 5 gal. of purchased beef stock costs $40, a food service establishment can save $___ by preparing the stock in-house.

_______________ **33.** If 5 gal. of purchased chicken stock costs $41.50, a food service establishment can save $___ by preparing the stock in-house.

Activity: Matching Butter and Contemporary Sauces

Many chefs consider a sauce to be the unifying component of a dish. Butter and contemporary sauces often accompany a food to enhance the flavor and/or moistness of a dish. Sauces can also add a variety of textures and enliven the color of a finished dish.

Match each of the following butter and contemporary sauces with the correct description.

___________	**1.** Salsa or relish	___________	**6.** Pesto
___________	**2.** Coulis	___________	**7.** Nage
___________	**3.** Beurre blanc	___________	**8.** Foam
___________	**4.** Flavored oil	___________	**9.** Chutney
___________	**5.** Compound butter	___________	**10.** Broken butter

A. An aromatic court bouillon that is often used as a finishing sauce

B. A sauce made from fresh ingredients, such as cilantro, basil, mint, pine nuts, and parmesan cheese, that have been either crushed with a mortar and pestle or finely chopped in a food processor before mixing with olive oil

C. A sauce made from ingredients that are cooked with sugar and spices to yield a sweet-and-sour flavor

D. Produced by making a reduction of a flavoring ingredient, shallots, garlic, and wine; lecithin powder is added and the mixture is puréed

E. A sauce that can be served either warm or cold over grilled or sautéed items, as well as desserts

F. A flavorful butter sauce made by mixing cold, softened butter with flavoring ingredients such as fresh herbs, garlic, vegetable purées, dried fruits, preserves, or wine reductions

G. A butter sauce made by heating butter until the fat, milk solids, and water separate or "break"

H. A sauce made by mixing diced vegetables or fruits, herbs, and spices together; textures range from coarse to puréed

I. A butter-based emulsified sauce made by whisking cold, softened butter into a wine, white-wine vinegar, shallot, and peppercorn reduction

J. Often made from a neutral-flavored oil such as canola or grape seed oil

Chapter 9 Review
Soups

Section 9.1 Soup Varieties

Name ______________________________ **Date** ____________________

True-False

T F **1.** Soups are served hot, never chilled.

T F **2.** Soups can have a clear, thin consistency or a thick, velvety consistency.

T F **3.** Minestrone is a Vietnamese soup that contains beef or chicken, scallions, Welsh onions, charred ginger, wild coriander, basil, cinnamon, star anise, cloves, and black cardamom.

Multiple Choice

__________________ **1.** Minestrone is a soup from ___ that contains beans, onions, celery, carrots, stock, tomatoes, and optional pasta.

A. Spain
B. Germany
C. Italy
D. France

__________________ **2.** ___ is a soup from the Ukraine that contains beets, onions, beef stock or water, red-wine vinegar, dill, sugar, sour cream, and optional vegetables.

A. Gazpacho
B. Borscht
C. Goulash
D. Bouillabaisse

__________________ **3.** ___ soups include bisques, chowders, and cold soups, as well as unique regional varieties.

A. Specialty
B. Clear
C. Unique
D. Thin

Completion

__________________ **1.** Many soups are produced by using a high-quality ___ as a base and adding other ingredients to make the dish unique.

__________________ **2.** Soups can be made using fresh, seasonal ingredients or ___ leftovers.

__________________ **3.** Miso is a soup from ___ that contains fish broth, fermented soy, and dashi.

Name ______________________________ **Date** ______________

True-False

T F **1.** Allowing a broth to boil vigorously produces a broth of excellent quality.

T F **2.** To clarify a consommé, the clearmeat is mixed well and then whisked into the hot broth and brought to a low simmer.

T F **3.** A broth can be served as a finished product or made into a broth-based soup or consommé.

T F **4.** In a consommé preparation, once the clearmeat begins to form a solid mass, the mixture cannot be disturbed or stirred as the clearmeat could break apart and ruin the aroma of the consommé.

T F **5.** Although the clarification process adds a small bit of flavor, the bulk of a consommé's flavor is derived from the quality of the broth.

T F **6.** A consommé is a flavorful liquid made by simmering stock along with meat, poultry, seafood, or vegetables and seasonings.

T F **7.** After it is removed from a stockpot, a consommé does not need to be strained.

T F **8.** Stirring a consommé after a raft has formed can cause the final consommé to be cloudy.

Multiple Choice

______________ **1.** Consommés made from ___ should have a deep golden or amber color.
- A. beef
- B. veal
- C. pork
- D. poultry

______________ **2.** Beef broth is simmered for ___.
- A. 10–25 minutes
- B. 30–40 minutes
- C. 1 hour
- D. 2 hours or more

______________ **3.** Boiling a consommé can cause the final consommé to be ___.
- A. thick
- B. sticky
- C. cloudy
- D. clear

______________ **4.** A consommé is a very rich and flavorful ___ that has been further clarified to remove any impurities or particles that could cloud the finished product.
A. purée
B. roux
C. broth
D. velouté

Completion

______________ **1.** Fat droplets on the surface of a broth can be removed with a clean ___.

______________ **2.** A(n) ___ is a cold, ground, lean meat, fish, or poultry that is combined with an acid, ground mirepoix, and egg whites.

______________ **3.** ___ is a process that removes impurities, sediment, cloudiness, and particles from a liquid, such as stock.

______________ **4.** It is very important that a consommé be prepared in a nonreactive pot that is taller than it is ___.

______________ **5.** To perform a secondary clarification on a consommé, a mixture of lightly beaten ___, finely ground mirepoix, and either tomato purée or lemon juice is added to the consommé.

______________ **6.** A(n) ___ is the clearmeat that has risen to the surface of a consommé.

______________ **7.** A perfectly clear ___ shows that a chef has refined soup-making skills.

______________ **8.** After gently ___ for 1–2 hours, a consommé is removed through a hole in the raft using a ladle or from the bottom of a stockpot through the drain spigot.

Matching

______________ **1.** Broth

______________ **2.** Raft

______________ **3.** Clearmeat

______________ **4.** Consommé

______________ **5.** Clear soup

______________ **6.** Oignon brûlé

A. Clearmeat that has risen to the surface

B. Very rich and flavorful broth that has been further clarified to remove any impurities or particles that could cloud the finished product

C. Flavorful liquid made by simmering stock along with meat, poultry, seafood, or vegetables and seasonings

D. Half an onion that is charred on the cut side

E. Cold, ground, lean meat, fish, or poultry that is combined with an acid (such as wine, lemon juice, or a tomato product), ground mirepoix, and egg whites

F. Stock-based soup with a thin, watery consistency

Chapter 9 Review
Soups

Section 9.3 Thick Soups

Name ______________________________ **Date** ____________________

True-False

T F **1.** Purée soups tend to further thicken if they are made in advance and stored.

T F **2.** Cream soups and purée soups are the two basic types of specialty soups.

T F **3.** The roux preparation method consists of sweating a white mirepoix and the flavoring ingredient in water.

T F **4.** The main difference between a cream soup and a purée soup is that most cream soups are thickened by an added starch.

T F **5.** Cream soups are generally thicker and coarser than purée soups and are seldom strained.

T F **6.** If a purée soup is too thin, a small amount of roux may be whisked in and the soup brought to a simmer.

T F **7.** Purée soups can have a completely smooth texture for a sophisticated look or a coarse texture for a more rustic look.

Multiple Choice

______________ **1.** Regardless of the preparation method chosen, the main ingredients of a cream soup need to be ___ until tender in order to be puréed smooth.
A. steamed
B. boiled
C. simmered
D. stewed

______________ **2.** The first step in the roux preparation method of a cream soup consists of sweating a(n) ___ and the flavoring ingredient in fat.
A. white matignon
B. white mirepoix
C. oignon brûlé
D. sachet d'épices

_______________ **3.** When making a purée soup, a portion of the main ingredient is often reserved and added to the purée to add ___ to the soup.

A. texture
B. weight
C. flavor
D. color

_______________ **4.** Purée soups are made by cooking ___ or dried legumes in a broth until tender.

A. shellfish
B. meats
C. herbs
D. vegetables

_______________ **5.** ___ soups are traditionally made using either the velouté preparation method or the roux preparation method.

A. Clear
B. Cream
C. Purée
D. Cold

Completion

_______________ **1.** Due to the slightly lower ___ content of vegetable purée soups, rice or potatoes are sometimes added to increase the texture and add body to the soup.

_______________ **2.** A(n) ___ is a white stock thickened with a roux.

_______________ **3.** A(n) ___ is a soup having a thick texture and consistency.

Chapter 9 Review
Soups

Section 9.4 Specialty Soups

Name ______________________________ **Date** ____________________

True-False

T F **1.** Many cold soups are made using fruit or vegetable juice.

T F **2.** Cold soups include traditional cold cream soups such as gazpacho, a cold potato and leek soup.

T F **3.** Well-known bisques include lobster bisque, clam bisque, and mussel bisque.

T F **4.** The primary difference between cream soups and chowders is that chowders are not puréed.

T F **5.** When preparing a bisque, the sachet d'épices is not removed before the soup is puréed slightly.

T F **6.** The name bisque traditionally applies to a soup made from shellfish and thickened with cooked, puréed rice.

Multiple Choice

______________ **1.** A ___ is considered a specialty soup.
- A. consommé
- B. chowder
- C. purée
- D. broth

______________ **2.** Today, bisques are commonly thickened with a ___, which produces a bisque with a smoother and richer consistency.
- A. cooked rice
- B. roux
- C. cream-based liquid
- D. broth or consommé

______________ **3.** The majority of cream-based chowders are thickened with a ___.
- A. roux
- B. cooked rice
- C. beurre manié
- D. broth

______________ **4.** There are two basic types of ___ soups: those that require cooking the main ingredients and those that use raw ingredients that have been puréed.
A. cream
B. clear
C. thick
D. cold

Completion

______________ **1.** A bisque is a type of cream soup that is typically made from ___.

______________ **2.** A(n) ___ is hearty soup that contains visibly large chunks of the main ingredients.

______________ **3.** Although most cream-based chowders are thick, a few chowders are ___-based and therefore thinner.

______________ **4.** Bisques commonly include ___ for added richness and flavoring ingredients such as brandy or cognac for depth of flavor.

______________ **5.** To prepare gazpacho, fresh raw vegetables are puréed with tomato juice and spices and served ___.

Name ______________________________ Date ________________

Activity: Comparing House-Made and Convenience Consommés

A consommé is a very rich and flavorful broth that has been further clarified to remove any impurities or particles that could cloud the finished product. Preparing a consommé can be labor intensive and foodservice operations sometimes rely on ready-to-use convenience products.

Obtain a purchased, convenience beef consommé. Cost the following beef consommé recipe using the recipe costing form to answer the questions that follow.

Recipe Costing Form								
Recipe: Beef Consommé								
Yield: 1 gal.								
Portion Size: 8 oz								
Item No.	**Ingredient**	**Quantity**	**As-Purchased (AP)**		**Yield Percentage**	**Edible-Portion (EP)**		**Total Cost**
			Amount	**Cost**		**Amount**	**Cost**	
888	egg whites	8	18	$2.88	50%			
427	ground beef	3 lb	50 lb	$99.50	100%			
	mirepoix							
1148	medium onion, diced	8 oz	50 lb	$75.00	83%			
1120	celery, diced	4 oz	25 lb	$28.75	85%			
1162	carrot, diced	4 oz	25 lb	$28.75	92%			
1105	tomatoes, seeded and diced	2	25 lb	$72.00	90%			
	beef stock	5 qt	594 fl oz	$37.35	100%			
	oingon brûlé							
1148	medium onion	½	50 lb	$75.00	83%			
1822	bay leaf	3	8 oz	$5.75	100%			
	cloves	3	8 oz	$6.98	100%			
1886	salt	to taste	25 lb	$30.00	100%			
1822	bay leaf	3	8 oz	$5.75	100%			
1856	parsley	4 sprigs	8 oz	$1.50	96%			
1832	thyme	2 sprigs	8 oz	$6.88	100%			
1889	peppercorns	½ tsp	5 lb	$40.00	100%			
							Total	

________________ **1.** Based on the total cost of the beef consommé recipe, an 8 oz serving costs $___ to prepare.

________________ **2.** The total cost of the convenience consommé is ___.

____________________ **3.** The cost for an 8 oz serving of the convenience consommé is ___.

4. How does the beef consommé recipe compare to the convenience consommé in price?

____________________ **5.** The beef consommé recipe contains ___ calories per 8 oz serving.

____________________ **6.** The convenience consommé contains ___ calories per 8 oz serving.

____________________ **7.** The beef consommé recipe contains ___ g of total fat per 8 oz serving.

____________________ **8.** The convenience consommé contains ___ g of total fat per 8 oz serving.

____________________ **9.** The beef consommé recipe contains ___ mg of sodium per 8 oz serving.

____________________ **10.** The convenience consommé contains ___ mg of sodium per 8 oz serving.

11. Compare the nutrient values for the beef consommé recipe and the convenience consommé.

Prepare the beef consommé recipe and the convenience consommé.

12. How does the beef consommé recipe compare to the convenience consommé in taste?

Activity: Comparing Puréeing Methods

Puréed soups are prepared with a starchy vegetable such as potatoes, squashes, turnips, carrots, or dried legumes that are cooked in broth until tender and then puréed to a smooth consistency. There are many methods a chef can use to purée a soup. These include the use of a food mill, blender, or food processor.

Prepare a purée of split pea soup. Before puréeing the soup, divide it into three portions. Purée the first portion in a food mill.

1. Describe the texture and consistency of the soup.

Purée the second portion in a blender.

2. Describe the texture and consistency of the soup.

Purée the third portion in a food processor.

3. Describe the texture and consistency of the soup.

Complete the recipe as directed for each of the three portions.

4. Compare and contrast the three soups. Which method of puréeing produced the best results?

5. Under what circumstances would puréeing soup with a food mill be most desirable?

6. Under what circumstances would puréeing soup with a blender be most desirable?

7. Under what circumstances would puréeing soup with a food processor be most desirable?

Activity: Comparing Soup Preparation Methods

Cream soups are traditionally made using one of two preparation methods. The first method uses a velouté, and the second uses a roux. Specialty soups such as bisques also have different types of preparation methods. Bisques are typically prepared using either a roux or a cream preparation method.

Describe the soup as follows.

1. Describe how the velouté method is used to prepare a cream soup.

2. Describe how the roux method is used to prepare a cream soup.

Prepare the cream of broccoli recipe (velouté method) and describe the soup.

3. Describe the appearance, texture, and flavor of the cream of broccoli soup made using the velouté preparation method.

Prepare the cream of broccoli recipe (roux method) and describe the soup as follows.

4. Describe the appearance, texture, and flavor of the cream of broccoli soup made using the roux preparation method.

5. Compare and contrast the cream of broccoli made using the velouté method with the cream of broccoli made using the roux method.

Respond to the following questions.

6. How is the roux method used to prepare a bisque?

7. How is the cream method used to prepare a bisque?

Prepare the shrimp bisque recipe using the roux preparation method. Describe the soup as follows.

8. Describe the appearance, texture, and flavor of the shrimp bisque made using the roux preparation method.

Prepare the shrimp bisque recipe using the cream preparation method. Describe the soup as follows.

9. Describe the appearance, texture, and flavor of the shrimp bisque made using the cream preparation method.

10. Compare and contrast the shrimp bisque made using the roux method and the shrimp bisque made using the cream method.

Activity: Researching Soups

Soups are versatile foods that can range from flavorful broths to hearty chowders. A variety of soups can be found in every culture. Often these soups have unique and interesting pasts detailing how they came into tradition.

Choose a specific soup or a category of soup and write an essay using the following outline.

I. Introduction
   A. Soup or category of soup
II. History of the soup
   A. Country of origin
   B. Chef or person attributed to creation of soup (if any)
   C. Economic influences
III. Contemporary use
   A. Countries where soup is commonly served
   B. Cultural influences
   C. Common variations
IV. Recipe
V. Conclusion

1. Prepare and present a report about a chosen soup using the outline above. Use visuals, such as pictures and food demonstrations, to enhance the report. Include a list of the sources used for the research project.

Chapter 10 Review
Sandwiches

Section 10.1 Sandwich Varieties

Name ______________________________ **Date** ______________

True-False

T F **1.** The use of flatbreads or thinly sliced whole-grain breads adds nutrients and flavor without adding a lot of fat and calories.

T F **2.** Sandwiches are one of the least versatile categories of food.

T F **3.** A cubano sandwich is filled with ham and pork that has been marinated in mojo sauce, dill pickles, and Swiss cheese.

Multiple Choice

______________ **1.** Grilling a chicken breast instead of ___ it provides a healthier chicken sandwich that is lower in fat and calories.
A. steaming
B. baking
C. poaching
D. frying

______________ **2.** A(n) ___ is a French roll filled with cooked tuna, red onions, hard cooked eggs, olives, capers, tomatoes, anchovies, and red wine vinegar.
A. arepa
B. muffuletta
C. pan bagnat
D. po'boy

______________ **3.** A ___ is a baguette typically filled with fried shrimp, oysters, or catfish.
A. muffuletta
B. Cubano
C. torta
D. po'boy

Completion

______________ **1.** ___ have become the symbol of a quick and tasty meal on the go.

______________ **2.** A(n) ___ is a Greek pita filled with roasted lamb or beef cut from a vertical spit with garnishes that may include lettuce, tomatoes, onions, cucumbers, and a yogurt-based tzatziki sauce.

_______________ **3.** ___ is a Middle Eastern dish consisting of a pita filled with a fried patty or ball of spiced chickpeas with lettuce, tomatoes, cucumbers, onions, and a tahini or yogurt-based sauce.

Chapter 10 Review
Sandwiches

Section 10.2 Sandwich Fundamentals

Name ________________________________ **Date** ________________

True-False

T F **1.** Mayonnaise is a cooked emulsion made by combining oil, egg yolks, and lemon juice or vinegar.

T F **2.** Gloves should always be worn when handling cheese to prevent mold growth.

T F **3.** A common mistake is slicing sandwich meats thin instead of thick.

T F **4.** The sandwich filling should always be at room temperature.

T F **5.** Vegetables and fruits can be used as the main filling for a sandwich.

T F **6.** The base determines the shape of the sandwich and adds flavor, color, and nutritional value.

Multiple Choice

______________ **1.** ___ is often used as a binding agent to hold sandwich fillings together, such as with egg salad, chicken salad, tuna salad, and potato salad.
A. Pâté
B. Milk
C. Mayonnaise
D. Butter

______________ **2.** A sandwich ___ is the main ingredient in a sandwich.
A. spread
B. filling
C. garnish
D. base

______________ **3.** Some meat varieties used in sandwiches are referred to as ___ meats or processed meats.
A. deli
B. bistro
C. expensive
D. inexpensive

______ **4.** The primary difference between salad dressing and mayonnaise is that salad dressing does not contain ___.
A. vinegar
B. lemon juice
C. oil
D. egg yolks

______ **5.** Other than ___ varieties, breads should be wrapped airtight to prevent them from becoming stale.
A. soft
B. crusty
C. white
D. wheat

______ **6.** Chefs will often whip a small amount of ___ into butter to lighten its texture and density before using it as a sandwich spread.
A. vinegar
B. yogurt
C. oil
D. water

______ **7.** ___-based meat substitution products that are lower in fat and in cholesterol than other meats are popular sandwich fillings.
A. Beef
B. Veal
C. Turkey
D. Duck

______ **8.** Hard-cooked eggs can be sliced and used as a sandwich ___.
A. spread
B. base
C. stabilizer
D. garnish

______ **9.** It is important to place the ___ fillings next to the bread.
A. driest
B. moistest
C. hottest
D. coldest

______ **10.** ___ offer lower fat, cholesterol, and calorie alternatives to other types of spreads.
A. Mayonnaises
B. Butters
C. Salad dressings
D. Purées

Completion

____________________ **1.** ___ is the most common and easy-to-handle sandwich base.

____________________ **2.** A(n) ___ is a long, rectangular loaf of sandwich bread with four square sides and a fine, dry texture.

____________________ **3.** A sandwich ___ is a slightly moist, flavorful substance that seals the pores of the bread and creates a thin moisture barrier.

____________________ **4.** A(n) ___ salad is a salad made by combining a main ingredient, often a protein, with a binding agent such as mayonnaise or yogurt and other flavoring ingredients.

____________________ **5.** Many sandwiches are named for their ___, such as hot dogs, Thai turkey wraps, and Italian sausage and peppers.

____________________ **6.** Spreadable cheeses, an olive tapenade, a spicy green olive jardinière, and jams are examples of ___ spreads that enhance sandwiches.

____________________ **7.** Most sandwiches consist of four main components: a(n) ___, a spread, one or more fillings, and one or more garnishes.

____________________ **8.** Examples of ___ include lettuces, tomato and onion slices, pickle slices or spears, olives, raw vegetables, and grilled peppers.

Matching

____________________ **1.** Filling

____________________ **2.** Garnish

____________________ **3.** Base

____________________ **4.** Spread

A. Main ingredient in a sandwich and is stacked, layered, or folded on top of the base

B. Edible packaging that holds the contents of a sandwich

C. Complementary food item that is served on or with a sandwich

D. Slightly moist, flavorful substance that seals the pores of the bread and creates a thin moisture barrier

Chapter 10 Review
Sandwiches

Section 10.3 Hot Sandwiches

Name ______________________________ **Date** ____________________

True-False

T F **1.** A hot wrap sandwich is made by adding a spread and uncooked fillings to a flatbread and then cooking it.

T F **2.** Hot closed sandwiches are rarely served with cold garnishes.

T F **3.** A patty melt is a popular hot open-faced sandwich.

T F **4.** Grilled sandwich fillings do not need to be thoroughly cooked prior to assembly.

T F **5.** Hot sandwiches can be grouped into five basic types: hot open-faced sandwiches, hot closed sandwiches, hot wrap sandwiches, grilled sandwiches, and tea sandwiches.

T F **6.** The Monte Cristo consists of two pieces of bread filled with cooked ham, turkey, and Swiss cheese that is then soaked in beaten eggs before it is pan-fried.

Multiple Choice

______________ **1.** A popular ___ sandwich is the chimichanga.

A. fried
B. grilled
C. hot closed
D. cold closed

______________ **2.** A hot ___ sandwich is often covered with a sauce and is usually eaten with a fork.

A. closed
B. open-faced
C. grilled
D. wrap

______________ **3.** The most common hot closed sandwich is the ___.

A. panini
B. chimichanga
C. pizza
D. hamburger

Completion

__________ **1.** The most popular hot ___ sandwich is one that is usually not thought of as a sandwich: the pizza.

__________ **2.** A(n) ___ is an Italian clamshell-style grill made specifically to cook grilled sandwiches.

__________ **3.** Common examples of hot ___ sandwiches are burritos, tacos, fajitas, and enchiladas.

__________ **4.** Two common ___ sandwiches are the grilled cheese and Reuben.

__________ **5.** A(n) ___ is a variety of hot sandwich that consists of a tortilla wrap filled with precooked meat and beans that is then fried.

Matching

__________ **1.** Fried sandwich

__________ **2.** Hot wrap sandwich

__________ **3.** Hot open-faced sandwich

__________ **4.** Hot closed sandwich

__________ **5.** Grilled sandwich

A. Sandwich made by placing one or more hot fillings between two pieces of bread or a split roll or bun

B. Sandwich made by adding a spread and precooked fillings to a flatbread and then cooking it

C. Sandwich that consists of precooked fillings placed within a closed or wrapped sandwich and then fried

D. Sandwich made by adding precooked filling or cheese to bread that has been buttered on the exterior and then heated after assembly

E. Sandwich consisting of one or two slices of fresh, toasted, or grilled bread, topped with one or more hot fillings, and covered with a sauce, gravy, or a melted cheese topping

SECTION 10.4 COLD SANDWICHES

Name ______________________________ **Date** ____________________

True-False

T F **1.** To begin to prepare a multidecker sandwich, one side of two toasted slices of bread are coated with a spread.

T F **2.** A cold wrap sandwich is typically wrapped in parchment or waxed paper and cut in half on the bias to reveal the filling.

T F **3.** Cold closed sandwiches are the quickest to prepare and the most commonly served.

Multiple Choice

______________ **1.** The crusts of ___ sandwiches are trimmed off to make them easy to eat without creating a lot of crumbs.
- A. tea
- B. wrap
- C. open-faced
- D. submarine

______________ **2.** Cold sandwiches can be grouped into ___ distinct types.
- A. three
- B. four
- C. five
- D. six

______________ **3.** The tea sandwich originated in ___ as a snack served with afternoon tea.
- A. France
- B. Norway
- C. England
- D. Sweden

Completion

______________ **1.** A(n) ___ sandwich consists of three pieces of bread, a spread, and at least two layers of garnishes and fillings.

________________ **2.** A(n) ___ sandwich is created by spreading cream cheese on softened cracker bread called lahvosh and then layering thin slices of meats, cheeses, lettuce, and pickles before rolling it up like a jelly roll.

________________ **3.** A(n) ___ sandwich is often cut into rectangles, squares, triangles, or circles.

Matching

________________ **1.** Cold wrap sandwich

________________ **2.** Cold closed sandwich

________________ **3.** Tea sandwich

________________ **4.** Cold open-faced sandwich

A. Cold sandwich that consists of a single slice of bread that is often toasted or grilled and then coated with a spread and topped with fillings

B. Cold sandwich that consists of two pieces of bread, or the top and bottom of a bun or roll, coated with a spread and topped with one or more fillings and garnishes

C. Petite and delicate sandwich with a trimmed crust and soft filling

D. Cold sandwich in which a flat bread or tortilla is coated with a spread, topped with one or more fillings and garnishes, and rolled tightly

SECTION 10.5 SANDWICH STATIONS

Name ______________________________ **Date** ________________

True-False

T F **1.** The equipment needs of a sandwich station are determined by the menu.

T F **2.** Hand tools such as knives, cutting boards, spatulas, and spreaders are basic necessities for a sandwich station.

T F **3.** All foodservice operations require gloves during food handling.

T F **4.** The primary responsibility of a pastry chef is to assemble sandwiches quickly, neatly, consistently, and efficiently.

T F **5.** Sandwich makers may choose to wear hair restraints, which would exceed the minimum sanitation requirements.

Multiple Choice

_______________ **1.** When preparing sandwiches, all ingredients must be kept out of the ___ at all times.
A. chef's range of motion
B. refrigerator
C. temperature danger zone
D. sandwich station mise en place

_______________ **2.** When gloves are used, it is imperative that the gloves be ___ to avoid cross-contamination.
A. changed frequently
B. labeled with the employees initials
C. washed between sandwiches
D. oversized

_______________ **3.** All produce needs to be ___ before starting sandwich preparation.
A. wrapped in plastic
B. chemically treated
C. wilted
D. washed and dried

Completion

_______________ **1.** All meats should be sliced and portioned by ___.

_______________ **2.** Practicing proper ___ control helps maintain accurate food costs.

_______________ **3.** It is important to wash hands both before and after wearing ___.

SECTION 10.6 SANDWICH ASSEMBLY AND PLATING

Name ______________________________ **Date** ______________

True-False

T F **1.** Right-handed sandwich makers should place the bread supply to their right.

T F **2.** To begin to prepare sandwiches in quantity, bread or toast slices are arranged in rows on a sheet pan.

T F **3.** Hot sandwiches such as a cheeseburger or pulled pork sandwich are often served open-faced.

Multiple Choice

______________ **1.** A sandwich maker can determine a comfortable ___ by extending both arms directly out in front and then sweeping each arm in an arc to each side.
- A. personal space
- B. range of motion
- C. uniform size
- D. interpersonal distance

______________ **2.** Spreads and filling ingredients should be placed ___.
- A. in front of the sandwich maker
- B. to the right of the sandwich maker
- C. to the left of the sandwich maker
- D. below the work counter

______________ **3.** The ___ used on many hot closed sandwiches bind the bread to the filling.
- A. seasonings
- B. garnishes
- C. meats
- D. cheeses

______________ **4.** Cold sandwiches are typically cut into ___ to expose the interior of the sandwich and to make them easier to eat.
- A. slices
- B. unique shapes
- C. halves or quarters
- D. cubes

______ **5.** Frill pricks are quite visible to prevent ___.
A. a choking accident
B. the sandwich from falling apart
C. misplacing them in the kitchen
D. cross-contamination

Completion

______ **1.** Cold closed sandwiches are commonly held together with ___, which are slightly longer toothpicks that have a frilly plastic decoration at one end and a sharp point on the other end.

______ **2.** Having a paper ___ around a sandwich not only keeps the hands cleaner but is a fast, safe, and convenient way to hold the entire sandwich together.

______ **3.** Cheeses, wrappers, and frill picks serve as ___ that secure sandwich components together.

______ **4.** Common ___ such as coleslaw, pasta salad, french fries, or cut fruit add visual appeal, flavor, and nutrition to plated sandwiches.

______ **5.** All of the ingredients in a sandwich station should be within the range of ___ to maximize efficiency and speed.

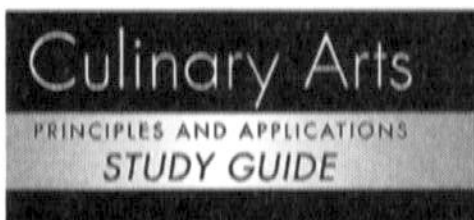

Name ______________________________ Date ________________

Activity: Creating Healthful Sandwiches

As a part of menu planning, a chef must provide options for customers desiring healthful menu options. Substituting ingredients with healthful ones, reducing the size or number of less healthy ingredients, or changing the method of preparation can create a healthful menu option.

Analyze each of the sandwiches listed and then answer the questions that follow.

Turkey Club Sandwich

1. What are three changes that will make a turkey club sandwich healthful?

2. Why will these changes make this turkey club sandwich healthful?

Chicken Salad Sandwich

3. What are three changes that will make a chicken salad sandwich healthful?

4. Why will these changes make this chicken salad sandwich healthful?

Cheeseburger and French Fries

5. What are three changes that will make a cheeseburger and French fries combination healthful?

6. Why will these changes make this cheeseburger and French fries combination healthful?

Monte Cristo Sandwich

7. What are three changes that will make a Monte Cristo sandwich healthful?

8. Why will these changes make this Monte Cristo sandwich healthful?

Activity: Researching Sandwiches

Sandwiches are one of the most versatile categories of food and allow chefs to showcase their imagination. Special ingredient combinations are used to create signature sandwiches around the world.

Choose a specific sandwich and write an essay using the following outline.

I. Introduction
  A. Sandwich type
II. History of the sandwich
  A. Country of origin
  B. Chef or person attributed to creation of sandwich (if any)
  C. Economic influences
III. Contemporary use
  A. Country/countries where the sandwich is commonly served
  B. Cultural influences
  C. Common variations
IV. Recipe
V. Conclusion

1. Prepare and present a report about a chosen sandwich using the outline above. Use visuals to enhance the report, such as pictures and food demonstrations. Include a list of sources used for the research project.

Activity: Identifying Sandwich Components

Most sandwiches consist of four main components: a base, a spread, one or more fillings, and one or more garnishes. Understanding each of the main sandwich components helps a chef to choose interesting combinations that complement each other and deliver satisfying flavors and an enticing presentation.

Identify each of the following ingredients as a base, spread, filling, or garnish.

____________________ **1.** Cream cheese

____________________ **2.** Smoked salmon

____________________ **3.** Onion slice

____________________ **4.** Egg salad

____________________ **5.** Baguette

____________________ **6.** Nori

____________________ **7.** Sliced turkey

____________________ **8.** Mayonnaise

____________________ **9.** Lettuce leaf

____________________ **10.** Jam

____________________ **11.** Italian sausage

____________________ **12.** Tortilla

____________________ **13.** Pickle spear

____________________ **14.** Sauerkraut

____________________ **15.** Lime wedge

____________________ **16.** Pita

____________________ **17.** Compound butter

____________________ **18.** Hot pepper

____________________ **19.** Rice paper wrapper

____________________ **20.** Whole olive

Activity: Identifying Sandwich Types

Sandwiches are divided into hot and cold categories, depending on the serving temperature of the fillings. They are served open-faced, closed, or as wraps. Sandwiches also may be grilled, fried, or cut into various shapes before service.

Respond to the following.

1. Describe a cold closed sandwich.

2. Identify a cold closed sandwich from the recipes in chapter 10.

3. Describe a multidecker sandwich.

4. Identify a multidecker sandwich from the recipes in chapter 10.

5. Describe a hot open-faced sandwich.

6. Identify a hot open-faced sandwich from the recipes in chapter 10.

7. Describe a tea sandwich.

8. Identify a tea sandwich from the recipes in chapter 10.

9. Describe a cold wrap sandwich.

10. Identify a cold wrap sandwich from the recipes in chapter 10.

11. Describe a fried sandwich.

12. Identify a fried sandwich from the recipes in chapter 10.

13. Describe a hot closed sandwich.

14. Identify a hot closed sandwich from the recipes in chapter 10.

Chapter 11 Review
Eggs and Breakfast

Section 11.1 Egg Fundamentals

Name ______________________________ **Date** ______________

True-False

T F **1.** Although the yolk is only about one-third the weight of the egg, it contains more than three-fourths of the calories and one-fourth of the cholesterol found in the egg.

T F **2.** An egg held for even 10 minutes will show signs of quality loss.

T F **3.** As one of the eight most common food allergens, current FDA labeling regulations require that the presence of eggs be marked on food labels in bold print.

T F **4.** In addition to being served as a dish, eggs are also commonly used as a clarifying agent when preparing a consommé or aspic.

Multiple Choice

______________ **1.** The highest quality egg, Grade ___, has a firm yolk and white that both stand tall when the egg is broken onto a flat surface.
A. AA
B. A
C. BB
D. B

______________ **2.** Leftover egg whites and egg yolks should be tightly covered with plastic wrap before being refrigerated and should be used within two to ___ days.
A. three
B. four
C. five
D. six

______________ **3.** ___ are used almost exclusively to produce cholesterol free or lower cholesterol baked goods.
A. Whole eggs
B. Egg yolks
C. Eggless egg substitutes
D. Egg whites

______________ **4.** A large egg provides varying amounts of 13 essential nutrients, high-quality ___, and antioxidants that help fight cellular damage.
A. carbohydrates
B. fiber
C. cholestrol
D. protein

__________ **5.** Opened liquid-egg substitutes can be stored for ___.
A. only one day
B. three days
C. two weeks
D. six weeks

Completion

__________ **1.** The ___ is also known as the white.

__________ **2.** Eggs are classified according to size based on the minimum ___ per dozen eggs.

__________ **3.** In an egg, the ___ is the nutrition powerhouse because it contains the majority of the thiamin, pantothenic acid, folate, vitamins B6 and B12, calcium, iron, phosphorous, and zinc.

__________ **4.** A pasteurized egg is an egg that has been heated to a specific temperature for a specific period of time to kill ___ that can cause foodborne illnesses.

__________ **5.** An egg is composed of four main parts: the shell, shell membrane, yolk, and ___.

__________ **6.** ___ dates begin with January 1 as day 001 and number successively through December 31 as day 365 to represent the days of the calendar year.

Matching

__________ **1.** Albumen

__________ **2.** Chalazae

__________ **3.** Pasteurized egg

__________ **4.** Egg substitute

__________ **5.** Eggshell

__________ **6.** Eggless egg substitute

__________ **7.** Shell membrane

__________ **8.** Yolk

A. Egg that has been heated to a specific temperature for a specific period of time to kill bacteria that can cause foodborne illnesses

B. Clear portion of the raw egg, which makes up two-thirds of the egg and consists mostly of ovalbumin protein

C. Yellow-colored liquid composed of soy, vegetable gums, and starches derived from corn or flour

D. Thin, hard covering of an egg that is composed of calcium carbonate

E. Thin, skinlike material located directly under the eggshell

F. Liquid product that is typically made from a blend of egg whites, vegetable oil, food starch, powdered milk, artificial colorings, and additives

G. Yellow portion of the egg

H. Twisted cordlike strands that anchor the yolk to the center of the albumen

Name ______________________________ **Date** ________________

True-False

T F **1.** Scrambled eggs should always be slightly undercooked, because they become firm as they are held for service.

T F **2.** Peeling eggs in the shell is always begun at the narrow end of the egg while moving toward the large end.

T F **3.** A fried egg, scrambled eggs, or a mini omelet can be added to a hot biscuit, bagel, English muffin, tortilla, or croissant to create a delicious handheld breakfast.

T F **4.** A sunny-side up egg is a fried egg with an unbroken yolk that is cooked until the egg white has gone from translucent to white on the bottom and then flipped over and cooked on the other side.

T F **5.** When preparing poached eggs, oil is used to set the white firmly around the yolk when the egg is placed in the water.

T F **6.** A soufflé omelet is a lighter variation of a folded omelet.

Multiple Choice

______________ **1.** Sunny-side up, basted, over-easy, over-medium, and over-hard eggs are all referred to as ___ eggs.
A. poached
B. shirred
C. scrambled
D. fried

______________ **2.** The four types of ___ are rolled, folded, soufflé, and frittata.
A. omelets
B. egg sandwiches
C. poached eggs
D. shirred eggs

______________ **3.** A ___ egg is cooked in an oven at 350°F until the egg is set and cooked to the desired degree of doneness.
A. shirred
B. poached
C. fried
D. scrambled

______________ **4.** When baking a quiche, the custard should not be cooked to an internal temperature above ___°F.
A. 150
B. 175
C. 185
D. 200

______________ **5.** A rolled omelet is an omelet that is cooked and then rolled onto a plate and cooked filling ingredients are added ___.
A. on top of the omelet
B. through a slit cut into the top
C. through the ends
D. on the side and as a garnish

______________ **6.** Hard-cooked eggs are cooked for ___ minutes and have a bright, solid white and a pale, crumbly yolk.
A. 2–3
B. 4–6
C. 9–10
D. 12–15

Completion

______________ **1.** ___ is the easiest method to use when preparing eggs in quantity.

______________ **2.** Traditional eggs Benedict consists of two toasted English muffin halves, each topped with a slice of Canadian bacon and a(n) ___ egg.

______________ **3.** If fried eggs are cooked in butter, only ___ butter should be used due to it having a higher smoke point than whole butter, which burns easily.

______________ **4.** The yolk of an egg cooked in the shell may form a(n) ___ outer ring when overcooked due to the sulfur in the egg white reacting with the iron in the yolk.

______________ **5.** Eggs in the shell that are cooked for 3 minutes are considered ___-cooked.

______________ **6.** About ___ eggs can be poached at a time in a gallon of water.

Matching

______________ **1.** Omelet

______________ **2.** Quiche

______________ **3.** Shirred egg

______________ **4.** Frittata

______________ **5.** Over-medium egg

______________ **6.** Sunny-side up egg

A. Fried egg with a completely cooked white and a yolk that is cooked nearly all the way through

B. Egg dish made with beaten eggs that is cooked into a solid form

C. Egg that is baked on top of other ingredients in a shallow dish in the oven

D. Traditional folded omelet served open-faced after being browned in a broiler or hot oven

E. Lightly fried egg with an unbroken yolk that is not flipped over to cook the other side

F. Baked egg dish composed of a savory custard baked in a piecrust

Chapter 11 Review
Eggs and Breakfast

Section 11.3 Pancakes and Waffles

Name ______________________ **Date** ______________

True-False

T F **1.** Pancakes and waffles may be prepared in advance but doing this will sacrifice some quality.

T F **2.** Pancakes and waffles prepared in advance are warmed under a broiler or in a steamer before they are served.

T F **3.** Waffles are most often served with syrup, a fruit topping, and whipped cream.

T F **4.** Pancakes and waffles are rarely served with breakfast meats, such as bacon or sausage.

Multiple Choice

______________ **1.** A pancake is ready to be flipped over when the surface is nearly covered in ___.
A. dents
B. lines
C. bubbles
D. brown color

______________ **2.** The procedure for making waffles is very similar to ___ except that a waffle iron is used in place of a griddle or a pan.
A. pancakes
B. blintzes
C. doughnuts
D. fritters

______________ **3.** The basic recipe for pancakes consists of flour, sugar, salt, baking powder, baking soda, milk or buttermilk, ___, and butter or oil.
A. cinnamon
B. yeast
C. water
D. eggs

_______________ **4.** A ___ is used to cook waffle batter on both sides simultaneously.
A. sheet pan
B. waffle iron
C. griddle
D. sauté pan

Completion

_______________ **1.** ___ are also known as hotcakes, griddle cakes, or flapjacks.

_______________ **2.** Before warming precooked pancakes and waffles, they are ___-layered on sheet pans.

_______________ **3.** To prepare pancakes and waffles, the ___ ingredients are sifted together.

Chapter 11 Review
Eggs and Breakfast

Section 11.4 French Toast

Name ______________________________ **Date** ____________________

True-False

T F **1.** French toast is a popular breakfast item prepared by dipping slices of bread into a batter made from eggs, milk, sugar, and vanilla.

T F **2.** French toast is typically served with a fruit topping, syrup, or confectioner's sugar.

T F **3.** French toast and crunchy French toast can only be prepared with French bread.

Multiple Choice

______________ **1.** When making traditional French toast, the batter-dipped bread is placed on a lightly oiled ___ and cooked on each side until golden brown.
A. grill
B. waffle iron
C. sheet pan
D. griddle

______________ **2.** Crunchy French toast is cooked ___ until it is golden on both sides.
A. in a fryer
B. on a griddle or in a pan
C. in a steamer
D. with a waffle iron

Completion

______________ **1.** After crunchy French toast is cooked until golden brown on both sides, it is baked in a ___°F oven for 10 minutes until it is crunchy and cooked completely through.

______________ **2.** Using slightly ___ bread for crunchy French toast helps ensure that the outside is somewhat crisp and the inside is creamy but not soggy.

Chapter 11 Review
Eggs and Breakfast

Section 11.5 Crêpes and Blintzes

Name ______________________________ **Date** ____________________

True-False

T F **1.** Typically, crêpes are filled with fruit or cheese, rolled up, and topped with brown sugar.

T F **2.** The most common filling for a savory blintz is potato and onion.

T F **3.** Unlike crêpes, blintzes are only filled with savory ingredients.

Multiple Choice

______________ **1.** Crêpe batter does not contain ___ and therefore remains very thin when cooked.
- A. flour
- B. eggs
- C. leavening agents
- D. butter

______________ **2.** After blintzes are cooked on one side, filled, and then rolled, they are put in a hot ___ with clarified butter and browned on both sides.
- A. sauté pan
- B. oven
- C. fryer
- D. panini press

______________ **3.** Blintzes are rolled or folded into ___.
- A. triangles
- B. pillow-shaped parcels
- C. dumplings
- D. semicircles

Completion

______________ **1.** A(n) ___ is a French pancake that is light and very thin.

______________ **2.** A(n) ___ is a crêpe that is only cooked on one side and not flipped over to cook the other side.

Name ______________________________ Date ________________

True-False

T F **1.** Breakfast steaks and pork chops are usually cut thick for a shorter cooking time.

T F **2.** Pork bacon is the least common breakfast bacon.

T F **3.** Turkey sausages offer a lower fat and calorie alternative to pork sausage.

T F **4.** Sausage links are typically 4 oz to 5 oz each.

T F **5.** The most common thickness of sliced bacon is an 18–22 count, meaning there will be between 18 and 22 slices per pound.

Multiple Choice

______________ **1.** Lox is a brine-cured ___ that has been cold smoked.
A. herring
B. salmon
C. ham
D. bacon

______________ **2.** Breakfast sausages are served in the form of ___.
A. cakes or links
B. cakes or sticks
C. patties or links
D. patties or sticks

______________ **3.** ___ bacon is packaged with each slice separated and laid out on sheets of parchment paper.
A. Slab-packed
B. Layer-packed
C. Shingled
D. Offset

______________ **4.** ___ is a Polish pork or beef sausage flavored with garlic, pimento, and cloves.
A. Chorizo
B. Kielbasa
C. Boudin
D. Bratwurst

______________ **5.** ___ is a highly seasoned Creole link sausage made of pork, pork liver, and rice.
A. Boudin
B. Kielbasa
C. Chorizo
D. Bratwurst

Completion

______________ **1.** ___ is shredded and chopped meat that has been mixed and cooked with diced potatoes, onions, and seasonings.

______________ **2.** ___ is an unsmoked pork belly that has been cured in salt and spices, such as nutmeg and pepper, and then dried for about three months.

______________ **3.** ___ is a hamlike breakfast meat made from boneless, smoked, pressed pork loin.

______________ **4.** A(n) ___ is a whole herring that has been split from tail to head, gutted, salted or pickled, and cold smoked.

______________ **5.** Chicken-fried steak is also known as ___-fried steak.

Name ______________________________ **Date** ________________

True-False

T F **1.** Although the name "hash brown" refers to a chopped, recooked potato, most people refer to the shredded potato dish as hash browns.

T F **2.** Boiling the dough in a salt water solution prior to baking is what gives bagels their tough, chewy exterior.

T F **3.** A fritter, also known as a beignet, is a fried donutlike item that may or may not be filled with fruit.

T F **4.** Muesli is a baked mixture of rolled oats, nuts, dried fruit, and honey.

T F **5.** Frozen or canned fruit is typically used to fill crêpes and blintzes and top pancakes, waffles, or French toast.

Multiple Choice

______________ **1.** Breakfast potatoes are usually served ___.
A. fried
B. baked
C. mashed
D. steamed

______________ **2.** ___ are boiled until tender, cooled and diced or sliced, and then browned on all sides in a hot sauté pan.
A. Potato pancakes
B. Hash browns
C. Home fries
D. French fries

______________ **3.** ___ are often used as the base for breakfast sandwiches.
A. Sweet muffins
B. Fritters
C. Quick bread loaves
D. English muffins

______________ **4.** A yogurt is a tangy, custard-like cultured dairy product produced by adding a safe bacteria and ___ to milk.
A. sugar
B. an acid
C. an artificial sweetener
D. salt

______________ **5.** ___ are often topped with sausage gravy.
A. Biscuits
B. Bagels
C. Poppyseed muffins
D. Fritters

Completion

______________ **1.** It is important to note that yogurt has the same fat content as the type of ___ from which it is produced.

______________ **2.** A(n) ___ is a quick bread made by mixing solid fat, baking powder or baking soda, salt, and milk with flour.

______________ **3.** Sweet muffins are cup-shaped ___ breads that may be served with any meal.

______________ **4.** ___ is an unbaked mixture of rolled oats, wheat flakes, oat bran, raisins, dates, sunflower seeds, hazelnuts, and wheat germ.

______________ **5.** ___-style yogurt is made by straining most of the waterlike whey from the yogurt and adding milk solids during production.

Chapter 11 Review
Eggs and Breakfast

Section 11.8 Breakfast Cereals

Name ______________________________ **Date** ____________________

True-False

T F **1.** Oatmeal can be made from crushed, rolled, steel cut, or coarsely ground oats that are slowly simmered.

T F **2.** Farina is baked until the cornmeal becomes tender and creamy.

T F **3.** Cold cereals are typically served with a side of whole, low-fat, or fat-free milk, a side of fruit or toast, and juice.

T F **4.** Foodservice operations typically purchase prepackaged cereals in bulk or in single-serving sized boxes.

Multiple Choice

____________ **1.** ___ is commonly sold under the brand name of Cream of Wheat®.
A. Porridge
B. Oatmeal
C. Hominy
D. Farina

____________ **2.** Prepackaged breakfast cereals simply require the addition of ___.
A. water
B. milk
C. butter
D. yogurt

____________ **3.** Both white and yellow grits have a(n) ___ flavor and are often served with eggs and toast or biscuits.
A. earthy
B. sweet
C. salty
D. buttery

____________ **4.** Hot cereals need to be cooked on the stovetop or in a ___.
A. ramekin
B. bain marie
C. steam-jacketed kettle
D. springform pan

Completion

____________________ **1.** Grits are made from ground corn called ___.

____________________ **2.** ___ is a hot breakfast dish made by heating a cereal grain in milk, water, or both.

____________________ **3.** Porridge, oatmeal, farina, and grits are popular hot breakfast cereals made from cracked, flaked, or ___ grains.

Chapter 11 Review
Eggs and Breakfast

Section 11.9 Breakfast Beverages

Name ______________________________ **Date** ____________________

True-False

T F **1.** Skim milk must contain less than 0.25% milk fat in order to be labeled as skim milk.

T F **2.** The grade of a tea refers to the quality of the drink produced by the leaves.

T F **3.** During processing, the pulp, parchment, and silverskin are removed from the coffee bean.

T F **4.** Decaffeinated coffee allows a coffee drinker to consume coffee without the effects of the stimulant.

T F **5.** A standard serving of juice is a 10 fl oz glass.

Multiple Choice

______________ **1.** Espresso is an intensely flavored coffee made from beans that have been roasted to the ___ stage.
A. medium
B. medium dark
C. dark
D. very dark

______________ **2.** Common ___ include black, green, oolong, and white.
A. coffees
B. teas
C. juices
D. milks

______________ **3.** ___ beans are typically used to produce higher-quality and more expensive coffees and are the best beans for brewing.
A. Canephora
B. Arabica
C. Robusta
D. Liberica

______________ **4.** Sencha is the most common variety of ___.
A. flavored milk
B. coffee
C. black tea
D. green tea

______________ **5.** Whole milk must contain at least ___% milk fat.
A. 3.5
B. 3.75
C. 4.0
D. 4.25

Completion

______________ **1.** The small red fruit of the coffee tree, known as a cherry, contains two small green seeds called coffee ___.

______________ **2.** A(n) ___ is a thick, blended drink that is typically made with fruit and/or vegetables and a liquid ingredient such as juice, milk, or a milklike beverage.

______________ **3.** A(n) ___ is an herbal beverage created by steeping herbs, spices, flowers, dried fruits, or roots in boiling water.

______________ **4.** ___ is the primary breakfast drink for a majority of the world's population.

______________ **5.** ___ teas undergo the least amount of processing.

Chapter 11 Review
Eggs and Breakfast

Section 11.10 Breakfast Plating Styles

Name ______________________________ **Date** ______________

True-False

T F **1.** Breakfast buffets allow a foodservice operation to charge a higher price for breakfast.

T F **2.** Skillet ingredients are typically cooked prior to being placed in the skillet.

T F **3.** Common plating of breakfasts includes placement of the main protein as the focal point of the dish on a square plate.

Multiple Choice

______________ **1.** Breakfast accompaniments such as toast are commonly served on side plates that are ___ the entrée plate.
A. larger than
B. the same size as
C. half the size of
D. three-quarters the size of

______________ **2.** Skillet breakfasts generally include eggs that are placed ___.
A. underneath other ingredients
B. in a ramekin
C. on a side plate
D. on top of other ingredients

______________ **3.** Guests perceive more value if their meal is provided on ___.
A. larger plates
B. multiple plates
C. yellow plates
D. skillets

Completion

______________ **1.** A(n) ___ is a light breakfast consisting of fruit, juice, toast, pastries, coffee, and tea.

______________ **2.** Common breakfast ___ include sliced or fanned strawberries and orange slices.

______________ **3.** Most people perceive breakfast ___ as a better value because there are more choices and no limit to how much a person can consume.

______________ **4.** After the ingredients are added, a hot ___ dish is finished in an oven or under a broiler.

Name ______________________________ **Date** ____________________

Activity: Analyzing Shell Egg Quality

Shell eggs should always be refrigerated at 40°F or below and may be kept refrigerated for about a month past the Julian date. The Julian date stamped on the egg carton indicates the day the eggs were packed. Julian dates begin with January 1 as day 001 and are numbered successively through December 31 as day 365 to represent the days of the calendar year.

Retrieve an egg that is more than 25 days past the Julian date on the carton. Break the egg carefully onto a plate.

1. What is the diameter of the egg white?

Retrieve an egg that is five to ten days past the Julian date on the carton. Break the egg carefully onto a plate.

2. What is the diameter of the egg white?

3. Why is there a difference in diameter, if any?

Activity: Preparing Eggs

Eggs must be cooked properly for a quality product. Eggs are used in culinary arts for many reasons including as an adhesive or to emulsify, thicken, bind, color, leaven, clarify, add moisture, or enrich other foods.

Boil an egg for 15 minutes. Allow the egg to cool to room temperature. Peel and slice the egg. Taste the egg.

1. How does the egg taste?

2. What is the texture of the egg?

Simmer an egg for 9 minutes. Run the egg under cold water to cool. Peel and slice the egg. Taste the egg.

3. How does the egg taste?

4. What is the texture of the second egg?

5. Compare and contrast the flavor and texture of the two eggs. Why is there a difference, if any?

Fry an egg in fat over high heat until the white is cooked and the center is set. Taste the egg.

6. How does the egg taste?

7. What is the texture of the egg?

Fry an egg over low heat until the white is cooked and the center is set. Taste the egg.

8. How does the egg taste?

9. What is the texture of the second egg?

10. Compare and contrast the flavor and texture of the two eggs. Why is there a difference, if any?

Poach an egg in simmering plain tap water until the egg is set. Taste the egg.

11. How does the egg taste?

12. What is the texture of the egg?

Poach an egg in simmering water with 1 tsp of salt and 1 tbsp of distilled white vinegar per quart of water until the egg is set. Taste the egg.

13. How does the egg taste?

14. What is the texture of the second egg?

15. Compare and contrast the flavor and texture of the two eggs. Why is there a difference, if any?

Prepare a rolled omelet without fillings and hold for 15 minutes. Taste the egg.

16. How does the egg taste?

17. What is the texture of the egg?

Prepare a rolled omelet without fillings. Taste the egg immediately.

18. How does the egg taste?

19. What is the texture of the egg?

20. Compare and contrast the flavor and texture of the two eggs. Why is there a difference, if any?

Activity: Calculating Breakfast Costs

Eggs and breakfast foods are very economical foods that provide good nutrition and appetizing entrées. When planning and pricing a breakfast menu, it is important to calculate total recipe cost and markup to set menu prices.

Calculate the cost of various breakfast items using the recipe costing form. Then complete the items below each form.

Recipe Costing Form

Recipe: Buttermilk Pancakes
Yield: 5 servings
Portion Size: 2 pancakes

Item No.	Ingredient	Quantity	As-Purchased (AP)		Yield %	Edible-Portion (EP)		Total Cost
			Amount	Cost		Amount	Cost	
1820	**All-purpose flour**	2 cups	50 lb	$60.00	100%			
1822	**Granulated sugar**	2 tbsp	50 lb	$49.10	100%			
1886	**Salt**	1 tsp	25 lb	$30.00	100%			
1833	**Baking powder**	2 tsp	5 lb	$9.71	100%			
1834	**Baking soda**	1 tsp	32 oz	$1.89	100%			
1760	**Buttermilk**	16 fl oz	½ gal.	$1.79	100%			
885	**Whole eggs (beaten)**	2	18	$2.88	100%			
840	**Unsalted butter (melted)**	2 tbsp	1 lb	$2.13	100%			
980	**Vegetable oil**	2 tbsp	1 gal.	$5.26	100%			
							Total	

_______________ **1.** The EP amount of all-purpose flour is ___.

_______________ **2.** The EP amount of granulated sugar is ___.

_______________ **3.** The EP amount of salt is ___.

_______________ **4.** The EP amount of baking powder is ___.

_______________ **5.** The EP amount of baking soda is ___.

_______________ **6.** The EP amount of buttermilk is ___.

_______________ **7.** The EP amount of whole eggs is ___.

_______________ **8.** The EP amount of unsalted butter is ___.

_______________ **9.** The EP amount of vegetable oil is ___.

_______________ **10.** The EP cost of all-purpose flour is $___ per lb.

_______________ **11.** The EP cost of granulated sugar is $___ per lb.

_______________ **12.** The EP cost of salt is $___ per lb.

_______________ **13.** The EP cost of baking powder is $___ per lb.

_______________ **14.** The EP cost of baking soda is $___ per oz.

_______________ **15.** The EP cost of buttermilk is $___ per gal.

_______________ **16.** The EP cost of whole eggs is $___ per egg.

_______________ **17.** The EP cost of unsalted butter is $___ per lb.

_______________ **18.** The EP cost of vegetable oil is $___ per gal.

_______________ **19.** If 1 cup of flour weighs 5 oz, the total cost of all-purpose flour is $___.

_______________ **20.** If 1 tbsp of sugar weighs 1 oz, the total cost of granulated sugar is $___.

_______________ **21.** If 1 tsp of salt weighs ½ oz, the total cost of salt is $___.

_______________ **22.** If 1 tsp of baking powder weighs ¼ oz, the total cost of baking powder is $___.

_______________ **23.** If 1 tsp of baking soda weighs ¼ oz, the total cost of baking soda is $___.

_______________ **24.** The total cost of buttermilk is $___.

_______________ **25.** The total cost of whole eggs is $___.

_______________ **26.** If there are 32 tbsp in 1 lb of butter, the total cost of unsalted butter is $___.

_______________ **27.** If there are 256 tbsp in a gal., the total cost of vegetable oil is $___.

_______________ **28.** The total cost for the buttermilk pancake recipe is $___.

Recipe Costing Form									
Recipe: Crêpes **Yield:** 8 servings									
Portion Size: 2 crêpes									
Item No.	**Ingredient**	**Quantity**	**As-Purchased (AP)**		**Yield %**	**Edible-Portion (EP)**		**Total Cost**
			Amount	**Cost**		**Amount**	**Cost**	
1750	**Whole milk**	1 qt	1 gal.	$2.89	100%			
1820	**All-purpose flour**	10 oz	50 lb	$60.00	100%			
885	**Whole eggs (beaten)**	6	18	$2.88	100%			
1886	**Salt**	½ tsp	25 lb	$30.00	100%			
840	**Unsalted butter (melted)**	3 oz	1 lb	$2.13	100%			
1822	**Granulated sugar**	3 oz	50 lb	$49.10	100%			
1826	**Confectioners' sugar**	4 oz	5 lb	$9.71	100%			
							Total	

_______________ **29.** The EP amount of whole milk is ___.

_______________ **30.** The EP amount of all-purpose flour is ___.

_______________ **31.** The EP amount of whole eggs is ___.

_______________ **32.** The EP amount of salt is ___.

_______________ **33.** The EP amount of unsalted butter is ___.

_______________ **34.** The EP amount of granulated sugar is ___.

_______________ **35.** The EP amount of confectioners' sugar is ___.

_______________ **36.** The EP cost of whole milk is $___ per gal.

_______________ **37.** The EP cost of all-purpose flour is $___ per lb.

_______________ **38.** The EP cost of whole eggs is $___ per egg.

_______________ **39.** The EP cost of salt is $___ per lb.

_______________ **40.** The EP cost of unsalted butter is $___ per lb.

_______________ **41.** The EP cost of granulated sugar is $___ per lb.

_______________ **42.** The EP cost of confectioners' sugar is $___ per lb.

_______________ **43.** The total cost of whole milk is $___.

_______________ **44.** The total cost of all-purpose flour is $___.

_______________ **45.** The total cost of whole eggs is $___.

_______________ **46.** If 1 tsp of salt weighs ½ oz, the total cost of salt is $___.

_______________ **47.** The total cost of unsalted butter is $___.

_______________ **48.** The total cost of granulated sugar is $___.

_______________ **49.** The total cost of confectioners' sugar is $___.

_______________ **50.** The total cost for the crêpe recipe is $___.

Recipe Costing Form									
Recipe: Waffles **Yield:** 5 servings									
Portion Size: 1 waffle									
Item No.	**Ingredient**	**Quantity**	**As-Purchased (AP)**		**Yield %**	**Edible-Portion (EP)**		**Total Cost**
			Amount	**Cost**		**Amount**	**Cost**	
885	**Whole eggs (beaten)**	2	18	$2.88	100%			
1750	**Whole milk**	12¾ fl oz	1 gal.	$2.89	100%			
1818	**Cake flour**	8½ oz	50 lb	$60.00	100%			
1833	**Baking powder**	1 tbsp	5 lb	$9.71	100%			
1822	**Granulated sugar**	3½ tbsp	50 lb	$49.10	100%			
840	**Unsalted butter (melted)**	3¼ oz	1 lb	$2.13	100%			
							Total	

_______________ **51.** The EP amount of whole eggs is ___.

_______________ **52.** The EP amount of whole milk is ___.

_______________ **53.** The EP amount of cake flour is ___.

_______________ **54.** The EP amount of baking powder is ___.

_______________ **55.** The EP amount of granulated sugar is ___.

_______________ **56.** The EP amount of unsalted butter is ___.

_______________ **57.** The EP cost of whole eggs is $___ per egg.

_______________ **58.** The EP cost of whole milk is $___ per gal.

_______________ **59.** The EP cost of cake flour is $___ per lb.

_______________ **60.** The EP cost of baking powder is $___ per lb.

_______________ **61.** The EP cost of granulated sugar is $___ per lb.

_______________ **62.** The EP cost of unsalted butter is $___ per lb.

_______________ **63.** The total cost of whole eggs is $___.

_______________ **64.** The total cost of whole milk is $___.

_______________ **65.** The total cost of cake flour is $___.

_______________ **66.** If 1 tsp of baking powder weighs ¾ oz, the total cost of baking powder is $___.

_______________ **67.** If 1 tbsp of sugar weighs 1 oz, the total cost of granulated sugar is $___.

_______________ **68.** The total cost of unsalted butter is $___.

_______________ **69.** The total cost for the waffle recipe is $___.

Recipe Costing Form

Recipe: French Toast
Yield: 4 servings
Portion Size: 2 pieces

Item No.	Ingredient	Quantity	As-Purchased (AP)		Yield %	Edible-Portion (EP)		Total Cost
			Amount	Cost		Amount	Cost	
1282	**White bread**	8 slices	20 slices	$1.25	100%			
885	**Whole eggs (beaten)**	3	18	$2.88	100%			
1750	**Whole milk**	⅔ cup	1 gal.	$2.89	100%			
1822	**Granulated sugar**	3¼ tsp	50 lb	$49.10	100%			
1855	**Vanilla extract**	½ tsp	1 qt	$9.75	100%			
1826	**Confectioners' sugar**	4 oz	5 lb	$9.71	100%			
							Total	

_______________ **70.** The EP amount of white bread is ___.

_______________ **71.** The EP amount of whole eggs is ___.

_______________ **72.** The EP amount of whole milk is ___.

_______________ **73.** The EP amount of granulated sugar is ___.

_______________ **74.** The EP amount of vanilla extract is ___.

_______________ **75.** The EP amount of confectioners' sugar is ___.

_______________ **76.** The EP cost of white bread is $___ per slice.

_______________ **77.** The EP cost of whole eggs is $___ per egg.

_______________ **78.** The EP cost of whole milk is $___ per gal.

_______________ **79.** The EP cost of granulated sugar is $___ per lb.

__________ **80.** The EP cost of vanilla extract is $___ per fl oz.

__________ **81.** The EP cost of confectioners' sugar is $___ per lb.

__________ **82.** The total cost of white bread is $___.

__________ **83.** The total cost of whole eggs is $___.

__________ **84.** The total cost of whole milk is $___.

__________ **85.** If 1 tsp of sugar weighs ⅓ oz, the total cost of granulated sugar is $___.

__________ **86.** If 1 tsp equals 0.17 fl oz, the total cost of vanilla extract is $___.

__________ **87.** The total cost of confectioners' sugar is $___.

__________ **88.** The total cost for the French toast recipe is $___.

Recipe Costing Form

Recipe: Spanish Omelet
Yield: 1 serving
Portion Size: 1 omelet

Item No.	Ingredient	Quantity	As-Purchased (AP)		Yield %	Edible-Portion (EP)		Total Cost
			Amount	Cost		Amount	Cost	
1148	**Medium onion, diced**	1 tsp	50 lb	$75.00	83%			
1126	**Red bell pepper, chopped**	1 tsp	25 lb	$35.99	85%			
1125	**Green bell pepper, chopped**	1 tsp	25 lb	$26.80	85%			
885	**Whole eggs (beaten)**	2	18	$2.88	100%			
980	**Vegetable oil**	1 tsp	1 gal	$5.26	100%			
1886	**Salt**	to taste	25 lb	$30.00	100%			
1890	**Pepper**	to taste	5 lb	$40.00	100%			
862	**Cheddar cheese, shredded**	1 tbsp	10 lb	$42.97	100%			
1320	**Prepared salsa**	1 tbsp	1 gal.	$24.00	100%			
1765	**Sour cream**	1 tsp	1 lb	$6.80	100%			
							Total	

__________ **89.** The EP amount of onion is ___.

__________ **90.** The EP amount of red bell pepper is ___.

__________ **91.** The EP amount of green bell pepper is ___.

__________ **92.** The EP amount of whole eggs is ___.

_______________ **93.** The EP amount of vegetable oil is ___.

_______________ **94.** The EP amount of salt is ___.

_______________ **95.** The EP amount of pepper is ___.

_______________ **96.** The EP amount of cheddar cheese is ___.

_______________ **97.** The EP amount of prepared salsa is ___.

_______________ **98.** The EP amount of sour cream is ___.

_______________ **99.** The EP cost of onion is $___ per lb.

_______________ **100.** The EP cost of red bell pepper is $___ per lb.

_______________ **101.** The EP cost of green bell pepper is $___ per lb.

_______________ **102.** The EP cost of whole eggs is $___ per egg.

_______________ **103.** The EP cost of vegetable oil is $___ per gal.

_______________ **104.** The EP cost of salt is $___ per lb.

_______________ **105.** The EP cost of pepper is $___ per lb.

_______________ **106.** The EP cost of cheddar cheese is $___ per lb.

_______________ **107.** The EP cost of prepared salsa is $___ per gal.

_______________ **108.** The EP cost of sour cream is $___ per lb.

_______________ **109.** If 1 tsp of onion weighs 1 oz, the total cost of onion is $___.

_______________ **110.** If 1 tsp of red bell pepper weighs 1 oz, the total cost of red bell pepper is $___.

_______________ **111.** If 1 tsp of green bell pepper weighs 1 oz, the total cost of green bell pepper is $___.

_______________ **112.** The total cost of whole eggs is $___.

_______________ **113.** If there are 256 tbsp in a gal., the total cost of vegetable oil is $___.

_______________ **114.** If ⅕ oz of salt is used to season to taste, the total cost of salt is $___.

_______________ **115.** If ⅕ oz of pepper is used to season to taste, the total cost of pepper is $___.

_______________ **116.** If 1 tbsp of cheddar cheese weighs 2 oz, the total cost of cheddar cheese is $___.

_______________ **117.** If there are 256 tbsp in a gal., the total cost of prepared salsa is $___.

_______________ **118.** If 1 tsp of sour cream weighs 1.5 oz, the total cost of sour cream is $___.

_______________ **119.** The total cost for the Spanish omelet recipe is $___.

Activity: Calculating Breakfast Net Profits per Serving

A successful foodservice operation stays in business by making profits. One essential component of making a profit is effective menu pricing and knowing food costs.

Compare the total cost calculated for each menu item in Activity: Calculating Breakfast Costs to the listed menu price below. Then, answer the following questions.

Menu	
Buttermilk pancakes	$5.95
Crêpes	$6.99
Belgian waffle	$6.50
French toast	$5.99
Spanish omelet	$6.75

__________ **1.** The net profit for a serving of buttermilk pancakes if the buttermilk pancake recipe makes 5 servings is $___.

__________ **2.** The net profit for a serving of crêpes if the crêpes recipe makes 8 servings is $___.

__________ **3.** The net profit for a serving of waffles if the waffle recipe makes 5 servings is $___.

__________ **4.** The net profit for a serving of French toast if the French toast recipe makes 4 servings is $___.

__________ **5.** The net profit for a serving of Spanish omelet if the Spanish omelet recipe makes 1 serving is $___.

6. List other costs that may factor into the preparation of menu items.

Chapter 12 Review
Fruits

Section 12.1 Types of Fruit

Name ______________________ **Date** ______________

True-False

T F **1.** Quinces are not eaten raw because they have a bitter taste.

T F **2.** A date is a plump, juicy, and meaty drupe that grows on a date palm tree.

T F **3.** A Comice pear is a small pear that is sometimes called a honey pear or sugar pear because of its syrupy, fine-grained flesh and complex sweetness.

T F **4.** A supreme is the flesh from a segment of a citrus fruit that has been cut away from the membrane.

T F **5.** A Satsuma is a small, seedless variety of lemon.

T F **6.** A Crenshaw melon is a teardrop-shaped melon with a thick, bright-yellow, ridged rind and white flesh.

T F **7.** Fruits are nutritious because they are high in water, dietary fiber, vitamins, and antioxidants.

T F **8.** A durian is an exotic fruit that contains several pods of sweet, yellow flesh and has a custard-like texture.

T F **9.** Lychees must be harvested ripe, as they do not continue to ripen after being harvested.

T F **10.** All varieties of apples are suitable for cooking purposes.

T F **11.** Fresh or frozen cranberries are always cooked with sugar or simple syrup before being added to a dish in order to soften their tart flavor.

T F **12.** An aggregate fruit is a cluster of very large fruits.

T F **13.** A persimmon is a bright-orange drupe that grows on a tree and is similar in shape to a tomato.

T F **14.** A pome, also known as a stone fruit, is a type of fruit that contains one hard seed or pit.

T F **15.** The peel is not removed when preparing mangoes for use in a dish.

T F **16.** The hard outer rind of a melon can be netted, ribbed, or smooth in texture.

Multiple Choice

______ **1.** Tomatoes, cucumbers, eggplants, and sweet peppers are ___ commonly used in the professional kitchen.
A. drupes
B. pomes
C. exotic fruits
D. fruit-vegetables

______ **2.** A ___ is an oval fruit that has a smooth skin and grows on woody vines in large clusters.
A. grape
B. melon
C. drupe
D. pome

______ **3.** ___ is an odorless gas that a fruit emits as it ripens.
A. Carbon dioxide
B. Nitrogen gas
C. Ethylene gas
D. Carbon monoxide

______ **4.** Zest is the colored, outermost layer of the peel of a citrus fruit that contains a high concentration of ___.
A. minerals
B. oil
C. vitamins
D. water

______ **5.** When removing the core of an apple with a ___, the apple is first cut in half.
A. peeler
B. fruit corer
C. chef's knife
D. mandoline

______ **6.** The juicy ___ is the result of a peach mutation.
A. nectarine
B. apricot
C. persimmon
D. mango

______ **7.** Breadfruit is about the size of a small ___.
A. orange
B. cantaloupe
C. grape
D. apple

______________ **8.** ___ cherries include Bing cherries, Gean cherries, and Rainier cherries.
A. Sweet
B. Sour
C. Light
D. Dark

______________ **9.** A pome is a fleshy fruit that contains a core of seeds and has a(n) ___ skin.
A. thick
B. ribbed
C. inedible
D. edible

______________ **10.** A ___ is eaten peel and all.
A. lime
B. kumquat
C. cantaloupe
D. jackfruit

______________ **11.** The texture of figs is somewhat tough and gritty due to the massive amount of tiny ___ inside.
A. pits
B. membranes
C. pods
D. seeds

______________ **12.** Watermelons are over ___% water.
A. 50
B. 75
C. 80
D. 90

Completion

______________ **1.** A(n) ___ is a fruit that is the result of crossbreeding two or more fruits of different species to obtain a completely new fruit.

______________ **2.** ___ grapes are commonly used in jams, jellies, and grape juice due to their high sugar content and distinguishable flavor.

______________ **3.** A quince is a hard yellow ___ that grows in warm climates.

______________ **4.** A(n) ___ is the thick outer rind of a citrus fruit.

______________ **5.** A(n) ___ is a tropical fruit that is a close relative of the banana but is larger and has a dark brown skin when ripe.

_______________ **6.** A Meyer lemon is a cross between a lemon and a(n) ___.

_______________ **7.** The rich, smooth, buttery flesh makes avocados a good sandwich ___ that can be used in place of mayonnaise, butter, or salad dressing.

_______________ **8.** A(n) ___ is a type of fruit that grows in the humid tropics where temperatures average 80°F with little temperature variation between winter and summer.

_______________ **9.** Dried plums are known as ___.

_______________ **10.** Ugli fruit is a hybrid made by crossbreeding a(n) ___ and a tangerine.

_______________ **11.** A(n) ___ grape is a seedless grape that is pale to light green in color.

_______________ **12.** Star fruit, mangosteens, durians, passion fruit, and kumquats are considered ___ fruits.

_______________ **13.** A(n) ___ is a pear- or cylinder-shaped tropical fruit weighing 1–2 pounds with flesh that ranges in color from orange to red-yellow.

_______________ **14.** Cantaloupes and ___ are often mistaken for one another in the marketplace.

_______________ **15.** A(n) ___ is a fruit that is the result of breeding two or more fruits of the same species that have different characteristics.

_______________ **16.** A(n) ___ is a type of fruit that is small and has many tiny, edible seeds.

_______________ **17.** A(n) ___ is a small, green or black drupe that is grown for both the fruit and its oil.

_______________ **18.** The ___ are the many flowers that wrap around a centralized core and develop into a pineapple.

SECTION 12.2 PURCHASING FRUIT

Name ______________________________ Date ____________________

True-False

T F **1.** Fresh fruit is packed in cartons, lugs, flats, crates, or bushels and is sold only by weight.

T F **2.** Left on the plant, fruit does not stop ripening when it reaches full maturity.

T F **3.** Canned fruit grades are based on a variety of characteristics including uniformity of shape, size, color, texture, and the absence of defects.

T F **4.** Fruits can be canned in water.

T F **5.** The ripening process can be accelerated by storing fruit at room temperature.

Multiple Choice

______________ **1.** ___ fruit has a much sweeter taste than fresh fruit due to the sugars being more concentrated.
- A. Frozen
- B. Irradiated
- C. Tropical
- D. Dried

______________ **2.** Apples, melons, and bananas give off ___ gas and should be stored away from delicate fruits.
- A. ethylene
- B. oxygen
- C. helium
- D. hydrogen

______________ **3.** Lugs can hold between ___ pounds of fruit.
- A. 25–50
- B. 35–60
- C. 50–75
- D. 75–100

__________ **4.** ___ is commonly used to produce a quick chill, which speeds the freezing process.
A. Ethylene gas
B. Liquid nitrogen
C. Carbon monoxide
D. Carbon dioxide

Completion

__________ **1.** ___ is the process of exposing food to low doses of gamma rays in order to destroy deadly organisms.

__________ **2.** Most fruit used in restaurants is either U.S. ___ or U.S. No. 1.

__________ **3.** ___ fruit has had most of the moisture removed.

__________ **4.** A(n) ___, the international symbol for irradiation, appears on the label of irradiated foods.

__________ **5.** Cans that are dented or bulging should be disposed of as they may contain harmful ___.

Name ______________________________ **Date** ________________

True-False

T F **1.** Adding sugar or lemon juice to fruit does not help prevent it from becoming mushy in the cooking process.

T F **2.** Poaching fruit is done at 185°F because the low temperature ensures that the fruit retains its shape while cooking.

T F **3.** Apples, bananas, pears, and peaches are suitable for frying because they do not break down when exposed to very high temperatures.

Multiple Choice

______________ **1.** Fruits that have a smaller amount of pectin include ___, raspberries, and peaches.
A. apples
B. cranberries
C. strawberries
D. grapes

______________ **2.** A(n) ___ is a fruit that is often poached.
A. kiwi
B. pear
C. cranberry
D. banana

______________ **3.** Fruit can be roasted at a high temperature in the oven to ___ their sugars.
A. clarify
B. coagulate
C. eliminate
D. caramelize

Completion

______________ **1.** The ___ method is often used to make stewed fruit and fruit compotes.

______________ **2.** ___ is a chemical present in all fruits that acts as a thickening agent when it is cooked in the presence of sugar and an acid.

______________ **3.** When broiling fruit, the fruit should be placed on a sheet pan lined with ___ paper.

Name ______________________ **Date** ______________

Activity: Identifying Fruits

Before purchasing or using fruit, it is essential to examine the fruit for freshness. The ability to identify fruits enables chefs to select high-quality products that result in dishes full of flavor and color.

Match each fruit to its image.

______ **1.** Plantains

______ **2.** Pomegranate

______ **3.** Lychee

______ **4.** Ugli fruit

______ **5.** Currants

______ **6.** Kiwifruit

______ **7.** Quinces

______ **8.** Grapes

______ **9.** Canary melon

______ **10.** Gooseberries

______ **11.** Muskmelon

______ **12.** Blackberries

______ **13.** Papaya

______ **14.** Nectarines

______ **15.** Mandarins

______ **16.** Dates

______ **17.** Durians

(A)

Frieda's Specialty Produce
(B)

Frieda's Specialty Produce
(C)

Frieda's Specialty Produce
(D)

(E)

Melissa's Produce
(F)

Frieda's Specialty Produce
(G)

Frieda's Specialty Produce
(H)

(I)

(J)

Frieda's Specialty Produce
(K)

Frieda's Specialty Produce
(L)

Frieda's Specialty Produce
(M)

(N)

Frieda's Specialty Produce
(O)

(P)

United States Department of Agriculture
(Q)

Activity: Classifying Fruits

Fruit comes in a variety of forms. There are several major classifications of fruit that can help indicate what the fruit will look like and how it will taste. These classifications include berries, grapes, pomes, drupes, melons, citrus fruits, tropical fruits, exotic fruits, and fruit-vegetables.

Classify each of the following fruits as berries, grapes, pomes, drupes, melons, citrus fruits, tropical fruits, exotic fruits, or fruit-vegetables. Letters can be used more than once.

______	**1.** Apricots	**A.** Berries
______	**2.** Passion fruit	**B.** Grapes
______	**3.** Tangerines	**C.** Pomes
______	**4.** Loganberries	**D.** Drupes
______	**5.** Pumpkin	**E.** Melons
______	**6.** Kumquats	**F.** Citrus fruits
______	**7.** Concord grapes	**G.** Tropical fruits
______	**8.** Watermelon	**H.** Exotic fruits
______	**9.** Currants	**I.** Fruit-vegetables
______	**10.** Grapefruits	
______	**11.** Star fruit	
______	**12.** Eggplants	
______	**13.** Apples	
______	**14.** Plums	
______	**15.** Lychees	
______	**16.** Pomegranates	
______	**17.** Boysenberries	
______	**18.** Figs	
______	**19.** Cherries	
______	**20.** Cantaloupes	
______	**21.** Mandarins	
______	**22.** Pears	
______	**23.** Mangoes	
______	**24.** Olives	
______	**25.** Tomatoes	

Activity: Analyzing Market Forms of Fruit

Frozen, canned, and dried fruit are often priced differently than fresh fruit. Regardless of whether a fruit is in season can also have an effect on its price. *Answers to questions 1–12 may vary.*

Choose a fruit and respond to the following items.

____________________ **1.** The fruit being priced is a(n) ___.

2. When is this fruit in season?

____________________ **3.** This fruit costs $___ per lb fresh, when in season.

____________________ **4.** This fruit costs $___ per lb fresh, when out of season.

____________________ **5.** This fruit costs $___ per lb frozen.

____________________ **6.** This fruit costs $___ per lb canned.

____________________ **7.** This fruit costs $___ per lb dried.

8. When would it be appropriate to substitute fresh fruit with frozen, canned, or dried fruit?

9. When would it not be appropriate to substitute fresh fruit with frozen, canned, or dried fruit?

10. How does processing this fruit affect its nutritional value?

11. In what form is this fruit the most nutritious?

12. In what form is this fruit the least nutritious?

Activity: Preparing Fruits

Chefs need to know proper knife techniques and procedures to effectively prepare fruits. With proper preparation, waste is reduced and safety is promoted.

Prepare the following fruits using proper knife skills. Have an instructor check the result and sign on the lines below when complete.

1. Core an apple.

Coring an Apple Proficiency

Instructor:____ Date:______________

2. Prepare an avocado.

Preparing an Avocado Proficiency

Instructor:____ Date:______________

3. Seed a melon.

Seeding a Melon Proficiency

Instructor:____ Date:______________

4. Cut a citrus fruit into supremes.

Cutting Citrus Fruit Supremes Proficiency

Instructor:____ Date:______________

5. Core a pineapple.

Coring a Pineapple Proficiency

Instructor:____ Date:_________________

6. Prepare a mango.

Preparing a Mango Proficiency

Instructor:____ Date:_________________

Activity: Cooking Fruit

There are many different ways to serve fruit. Common methods for cooking fruit include simmering, poaching, grilling, broiling, baking, roasting, sautéing, and frying. Each cooking method affects the appearance, texture, and flavor of the fruit.

Prepare the following recipe and respond to the items that follow.

Simmered or Poached Apples

1	apple
½ c	sugar
1½ c	water

1. Place sugar and water in a sauce pan and bring to a boil. Stir to dissolve sugar.
2. Wash, peel, core, and slice apple into ½ inch rings.
3. Add apple rings to hot syrup. Cover, reduce heat, and simmer for 20 minutes.

1. Describe the appearance, texture, and flavor of the simmered or poached apple.

2. Why does the simmered or poached apple have these characteristics?

Prepare the following recipe and respond to the items that follow.

Grilled or Broiled Apples

1	apple
2 tbsp	melted butter
to taste	cinnamon and sugar

1. Wash, peel, core, and slice apple into ½ inch rings.
2. Brush apple slices with melted butter and broil or grill until tender.
3. Sprinkle with cinnamon and sugar.

3. Describe the appearance, texture, and flavor of the grilled or broiled apple.

4. Why does the grilled or broiled apple have these characteristics?

Prepare the following recipe and respond to the items that follow.

Baked Apples

1	apple
1 tbsp	butter
2 tbsp	brown sugar
to taste	cinnamon
to taste	nutmeg

1. Wash apple.
2. Core apple with apple corer to within ½ inch of bottom end.
3. Place the apple upright in a custard cup and place butter in the hole where the core was removed.
4. Fill the center hole with brown sugar. Lightly dust the apple with cinnamon and nutmeg to taste.
5. Bake in a 400°F oven for 35 to 40 minutes or until soft, basting often with melted butter.

5. Describe the appearance, texture, and flavor of the baked apple.

6. Why does the baked apple have these characteristics?

Prepare the following recipe and respond to the items that follow.

Sautéed Apples

1	apple
2 tbsp	butter
2 tbsp	sugar
¼ tsp	cinnamon
dash	nutmeg

1. Wash, peel, and core apple. Slice as desired.
2. Melt butter in sauté pan.
3. Toss apple with sugar, cinnamon, and nutmeg.
4. Add apple to sauté pan and cook until a light golden brown.

7. Describe the appearance, texture, and flavor of the sautéed apple.

8. Why does the sautéed apple have these characteristics?

Prepare the following recipe and respond to the items that follow.

Fried Apple Rings

1	apple
1 c	vegetable oil
to taste	cinnamon and sugar

1. Wash, peel, core, and slice apple into ½ inch rings.
2. Submerge rings in hot fat and fry until tender.
3. Drain on paper towels. Sprinkle with cinnamon and sugar.

9. Describe the appearance, texture, and flavor of the fried apple rings.

10. Why does the fried apple have these characteristics?

Activity: Analyzing Enzymatic Browning Reactions

Many types of fruits can discolor or turn brown after they are sliced. Enzymatic browning, or oxidation, is the result of a chemical reaction that occurs when fruit enzymes are exposed to oxygen. Enzymatic browning is undesirable, and there are several methods used to prevent it.

Dip a freshly cut apple slice in ½ cup of a commercial ascorbic acid solution. Respond to the following items.

1. How long did it take for the apple slice to turn brown?

2. Describe the texture and flavor of the slice.

3. When would it be appropriate to use a commercial ascorbic acid solution to prevent browning?

Dip a freshly cut apple slice in ½ cup of lemon juice, pineapple juice, or orange juice. Respond to the following items.

4. How long did it take for the apple slice to turn brown?

5. Describe the texture and flavor of the slice.

6. When would it be appropriate to use lemon, pineapple, or orange juice to prevent browning?

Dip a freshly cut apple slice in ½ cup of a carbonated beverage. Respond to the following items.

7. How long did it take for the apple slice to turn brown?

8. Describe the texture and flavor of the slice.

9. When would it be appropriate to use a carbonated beverage to prevent browning?

Dip a freshly cut apple slice in ½ cup of sugar syrup. Respond to the following items.

10. How long did it take for the apple slice to turn brown?

11. Describe the texture and flavor of the slice.

12. When would it be appropriate to use sugar syrup to prevent browning?

Wrap a freshly cut apple slice tightly in plastic wrap. Respond to the following items.

13. How long did it take for the apple slice to turn brown?

14. Describe the texture and flavor of the slice.

15. When would it be appropriate to wrap a fruit tightly in plastic wrap to prevent browning?

Dip a freshly cut apple slice into 1 cup of boiling water for 30 seconds, remove promptly, and dip in ice water to blanch. Respond to the following items.

16. How long did it take for the apple slice to turn brown?

17. Describe the texture and flavor of the slice.

18. When would it be appropriate to blanch a fruit to prevent browning?

Chapter 13 Review
Vegetables

Section 13.1 Types of Vegetables

Name ______________________________ **Date** ______________________

True-False

T F **1.** It is important to clean leeks very well because soil and grit often become trapped between the layers of the bulb.

T F **2.** Although edible leaves, or greens, can be eaten raw, they are often cooked to decrease their bitterness.

T F **3.** Concassé is a preparation method in which a tomato is peeled, seeded, and then chopped or diced.

T F **4.** Fresh mushrooms should be firm and not spotted or slimy.

T F **5.** Summer squashes are harvested as mature vegetables one month after flowering.

T F **6.** Examples of edible roots include potatoes, sweet potatoes, yams, ocas, sunchokes, and water chestnuts.

T F **7.** An edible stem vegetable is the main trunk of a plant that develops buds and shoots instead of roots.

T F **8.** An eggplant is a deep-purple, white, or variegated fruit-vegetable with inedible skin and a yellow to white, spongy flesh.

T F **9.** Daikon radishes are stronger in flavor than red radishes.

T F **10.** The thin brown skin of a jicama must be removed before use.

T F **11.** The outer stalks of celery are sweeter and more tender than the inner stalks.

T F **12.** Examples of edible flowers include asparagus, celery, fennel, rhubarb, kohlrabi, and hearts of palm.

Multiple Choice

______________ **1.** A rutabaga is a round root vegetable derived from a cross between a Savoy cabbage and a ___.

A. beet
B. parsnip
C. celeriac
D. turnip

_______________ **2.** Ideal squash blossoms are ___ buds.
A. closed
B. open
C. small
D. brown

_______________ **3.** Okra is a green fruit-vegetable pod that contains small, round, ___ seeds and a gelatinous liquid.
A. black
B. white
C. green
D. yellow

_______________ **4.** A(n) ___ mushroom, also known as a forest mushroom, is an amber, tan, brown, or dark-brown mushroom with an umbrella shape and curled edges.
A. shiitake
B. morel
C. wood ear
D. enokitake

_______________ **5.** ___ has a mild, sweet flavor that is often associated with licorice or anise.
A. Chard
B. Rhubarb
C. Fennel
D. Kohlrabi

_______________ **6.** A shallot is a very small bulb vegetable that is similar in shape to garlic and has ___ cloves inside.
A. two or three
B. three or four
C. four or five
D. five or six

_______________ **7.** Watercress is a small, crisp, dark-green, edible leaf that is a member of the ___ family.
A. beet
B. chicory
C. cabbage
D. mustard

_______________ **8.** Some fresh bean and pea varieties are called edible ___, meaning that both the exterior skin and the interior seeds are edible.
A. stems
B. seeds
C. pods
D. roots

______________ **9.** ___ grow along an upright stalk and are ready to be harvested when they reach a diameter of about 1 inch.
A. Brussels sprouts
B. Head cabbages
C. Water chestnuts
D. Turnips

______________ **10.** ___ cabbages lack the sulfurlike odor often associated with cooking other cabbage varieties.
A. Head
B. Purple
C. Red
D. Savoy

______________ **11.** A lentil is a very small, dried ___ that has been split in half.
A. bean
B. pea
C. pulse
D. sprout

______________ **12.** ___ has a chewy texture and a savory flavor that is similar to bacon.
A. Nori
B. Arame
C. Wakame
D. Dulse

Completion

______________ **1.** A(n) ___ is the underwater root vegetable of an Asian water lily that looks like a solid-link chain about 3 inches in diameter and up to 4 feet in length.

______________ **2.** ___ is a tart stem vegetable that ranges in color from pink to red and is most often prepared like a fruit.

______________ **3.** ___ are green soybeans housed within a fibrous, inedible pod.

______________ **4.** A pulse is a dried seed of a(n) ___.

______________ **5.** Edible ___ include carrots, parsnips, salsify, radishes, turnips, rutabagas, beets, celeriac, jicamas, lotus roots, and bamboo shoots.

______________ **6.** A(n) ___ squash is a fruit-vegetable that grows on a vine and has a thick, hard, inedible skin and firm flesh surrounding a cavity filled with seeds.

______________ **7.** Dried ___ is ground to make celery salt.

______________ **8.** ___ is a potent compound that gives chiles their hot flavor.

______________ **9.** ___ grows covered with numerous layers of leaves attached to the stalk and surrounding the head to protect the head from sunlight and preserve its white color.

__________ **10.** The flavor of ___ is released when a clove is cut, crushed, or minced and increases the more finely the clove is cut.

__________ **11.** A(n) ___ mushroom is a trumpet-shaped mushroom that ranges in color from bright yellow to orange and has a nutty flavor and a chewy texture.

__________ **12.** There are thousands of varieties of ___, but the most popular varieties include beans, peas, pulses, and lentils.

Matching

__________ **1.** Edible leaf

__________ **2.** Legume

__________ **3.** Fruit-vegetable

__________ **4.** Edible root

__________ **5.** Edible seed

__________ **6.** Lentil

__________ **7.** Edible mushroom

__________ **8.** Edible flowers

__________ **9.** Vegetable

__________ **10.** Edible stem

__________ **11.** Sea vegetable

__________ **12.** Edible tuber

__________ **13.** Pulse

__________ **14.** Edible bulb

A. Seed of a nonwoody plant

B. Strongly flavored vegetable that grows underground and consists of a short stem base with one or more buds that are enclosed in overlapping membranes or leaves

C. Fleshy, spore-bearing body of an edible fungus that grows above the ground

D. Botanical fruit that is sold, prepared, and served as a vegetable

E. Main trunk of a plant that develops buds and shoots instead of roots

F. Flowers of nonwoody plants that are prepared as vegetables

G. Plant leaves that are often accompanied by edible leafstalks and shoots

H. Short, fleshy vegetable that grows underground and bears buds capable of producing new plants

I. Earthy-flavored vegetable that grows underground and has leaves that extend aboveground

J. Dried seed of a legume

K. Very small, dried pulse that has been split in half

L. Variety of edible saltwater plants that contain high amounts of dietary fiber, vitamins, and minerals

M. Edible root, bulb, tuber, stem, leaf, flower, or seed of a nonwoody plant

N. Edible seed of a nonwoody plant and grows in multiples within a pod

Chapter 13 Review
Vegetables

Section 13.2 Purchasing Vegetables

Name ______________________________ **Date** ______________

True-False

T F **1.** Frozen vegetables often retain their color and nutrients better than canned vegetables.

T F **2.** Dried onions and legumes are commonly used for convenience.

T F **3.** Potatoes, garlic, and onions should be stored in a refrigerator.

T F **4.** Recipes using fresh or slightly cooked vegetables require premium ingredients, while lesser grades are acceptable in soup recipes.

T F **5.** Canned vegetables are USDA graded as U.S. Prime, U.S. Choice, or U.S. Select.

Multiple Choice

______________ **1.** ___ vegetables can be stored for long periods if they are kept in a cool, dry place.
A. Frozen
B. Canned
C. Fresh
D. Cooked

______________ **2.** Fresh vegetables are less expensive during ___.
A. their off season
B. scarcity
C. their dormant period
D. their peak season

______________ **3.** Storing ___ in a refrigerator causes their starches to convert to sugars.
A. broccoli
B. potatoes
C. mushrooms
D. cucumbers

______________ **4.** Canned vegetables are packed by ___.
A. weight
B. volume
C. count
D. color

____________________ **5.** Frozen vegetables are usually packed in ___.
A. 10 oz or 16 oz boxes
B. 20 oz or 36 oz cartons
C. 1 lb or 2 lb bags
D. 6 lb crates or flats

Completion

____________________ **1.** Fresh vegetables are packed in cartons, lugs, flats, crates, or bushels and sold by weight or ___.

____________________ **2.** Most vegetables should be stored in a produce cooler at a temperature of ___°F or below.

____________________ **3.** ___ is the process of removing the water content from a food and replacing it with a gas.

____________________ **4.** Dried legumes need to be ___ before cooking.

____________________ **5.** Vegetables should always be stored away from fruits that emit ___, such as apples and bananas, as the gas can cause the vegetables to overripen and spoil.

SECTION 13.3 COOKING VEGETABLES

Name ______________________________ **Date** ________________

True-False

T F **1.** Fire-roasted vegetables cannot be cooked whole.

T F **2.** Grilling and broiling caramelize the sugars in vegetables, giving them a sweeter flavor.

T F **3.** Steamed vegetables can be finished by sautéing to add flavor.

Multiple Choice

________________ **1.** Vegetables such as onions, mushrooms, and zucchini are often batter-coated and ___ until crisp.
A. steamed
B. grilled
C. fried
D. broiled

________________ **2.** ___ is an organic pigment found in purple, dark-red, and white vegetables.
A. A carotenoid
B. A flavonoid
C. Chlorophyll
D. An alkali

________________ **3.** ___ is accomplished by placing fresh vegetables in boiling water and then quickly removing and placing them in an ice bath to stop the cooking process.
A. Blanching
B. Steaming
C. Roasting
D. Sweating

Completion

________________ **1.** The addition of acidic or alkaline ingredients when cooking vegetables causes chemical reactions that affect the ___ and texture of the vegetables.

________________ **2.** ___ is an organic pigment found in green vegetables.

________________ **3.** A carotenoid is an organic pigment found in ___ or yellow vegetables.

Name ______________________________ **Date** ____________________

Activity: Identifying Vegetables

A chef must be able to identify vegetables by appearance and smell. The appearance and smell of vegetables are also important in determining freshness and quality of ingredients.

Match each vegetable to its correct image.

______________ **1.** Shallots

______________ **2.** Leeks

______________ **3.** Butternut squash

______________ **4.** Cauliflower

______________ **5.** Bok choy

______________ **6.** Broccoli

______________ **7.** Brussels sprouts

______________ **8.** Chard

______________ **9.** Fennel

______________ **10.** Artichokes

______________ **11.** Eggplant

______________ **12.** Rutabaga

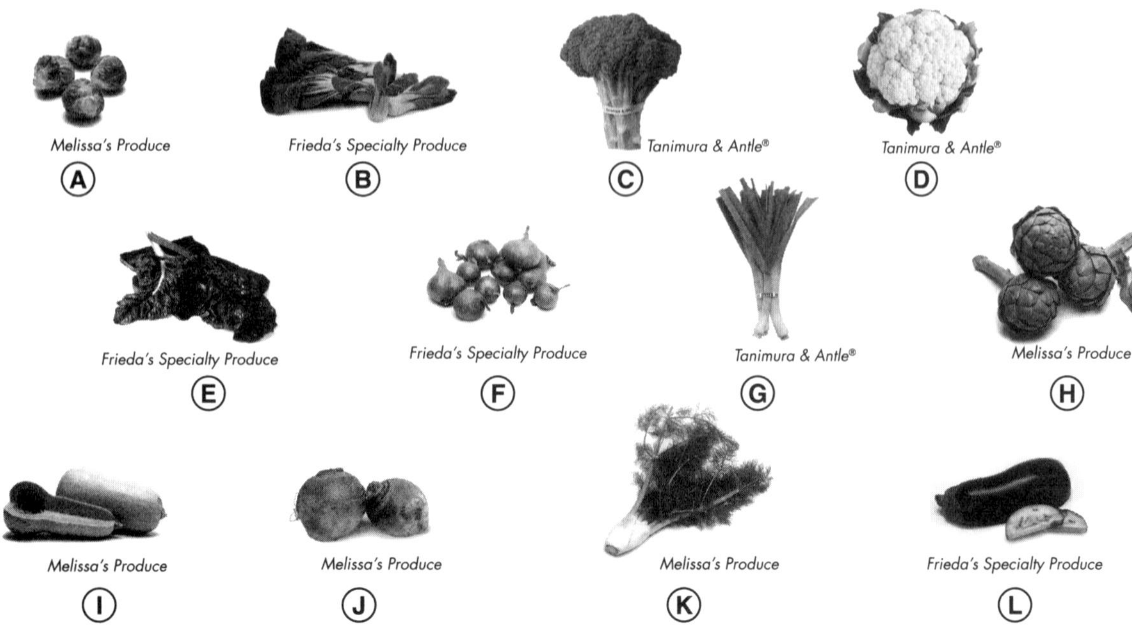

Activity: Identifying Vegetable Classifications

Vegetables come in many colors, shapes, sizes, and flavors and are essential to the creation of nutritious, colorful dishes. It is important to be able to identify the array of available vegetables and to select quality vegetables at their peak. Vegetables are classified as edible roots, bulbs, tubers, stems, leaves, flowers, or seeds of a nonwoody plant. Fruit-vegetables, sea vegetables, and mushrooms are typically prepared and served like vegetables.

Classify the following vegetables as roots, bulbs, tubers, stems, leaves, flowers, seeds of a nonwoody plant, fruit-vegetables, sea vegetables, or mushrooms. Letters can be used more than once.

__________ **1.** Broccoli

__________ **2.** Rhubarb

__________ **3.** Okra

__________ **4.** Garlic

__________ **5.** Sunchokes

__________ **6.** Black beans

__________ **7.** Bamboo shoots

__________ **8.** Jicama

__________ **9.** Wakame

__________ **10.** Watercress

__________ **11.** Yams

__________ **12.** Wax beans

__________ **13.** Artichokes

__________ **14.** Parsnips

__________ **15.** Porcini

__________ **16.** Hearts of palm

__________ **17.** Water chestnuts

__________ **18.** Acorn squash

__________ **19.** Nori

__________ **20.** Squash blossoms

__________ **21.** Chicory

__________ **22.** Sweet peppers

__________ **23.** Alfalfa sprouts

__________ **24.** Leeks

__________ **25.** Portobello

A. Roots

B. Bulbs

C. Tubers

D. Stems

E. Leaves

F. Flowers

G. Seeds

H. Fruit-vegetables

I. Sea vegetables

J. Mushrooms

Activity: Researching Legumes

A legume is the edible seed of a nonwoody plant and grows in multiples within a pod. There are thousands of varieties of legumes offering a rich source of fiber and protein with little or no fat.

Research one of the following types of legumes and respond to the items that follow.

- Beans
- Peas
- Pulses
- Lentils

1. Describe the chosen legume (variety, color, shape, size).

2. What is the origin of this legume?

3. Approximately when was this legume discovered?

4. How is this legume cultivated?

5. How is this legume commonly cooked?

6. Identify a recipe in which this legume is used.

Activity: Analyzing Market Forms of Vegetables

Substituting ingredients in a recipe can affect the cost, especially when the recipe is made in large quantities for a foodservice operation.

Choose a vegetable recipe. Record the recipe in the costing form and then respond to the following items.

1. Determine the cost of the recipe that includes fresh vegetables.

Recipe Costing Form - Using Fresh Vegetables								
Recipe:								
Yield:								
Portion Size:								
Item No.	Ingredient	Quantity	As-Purchased (AP)		Yield Percentage	Edible-Portion (EP)		Total Cost
			Amount	Cost		Amount	Cost	
							Total	

2. Determine the cost of the recipe if canned vegetables are used instead of fresh vegetables.

Recipe Costing Form - Using Canned Vegetables								
Recipe:								
Yield:								
Portion Size:								
Item No.	Ingredient	Quantity	As-Purchased (AP)		Yield Percentage	Edible-Portion (EP)		Total Cost
			Amount	Cost		Amount	Cost	
							Total	

3. Determine the cost of the recipe if frozen vegetables are used instead of fresh vegetables.

Recipe Costing Form - Using Frozen Vegetables

Recipe:

Yield:

Portion Size:

Item No.	Ingredient	Quantity	As-Purchased (AP)		Yield Percentage	Edible-Portion (EP)		Total Cost
			Amount	Cost		Amount	Cost	
							Total	

4. Which preparation is least expensive? Why?

Prepare the recipe three times. Use fresh vegetables in the first preparation, canned vegetables in the second preparation, and frozen vegetables in the third preparation. Respond to the following items.

5. Describe the appearance, texture, and flavor of the dish prepared with fresh vegetables.

6. Describe the appearance, texture, and flavor of the dish prepared with canned vegetables.

7. Describe the appearance, texture, and flavor of the dish prepared with frozen vegetables.

8. Which preparation produced the most appealing dish?

9. Which preparation would produce the most nutritious dish? Why?

Activity: Preparing Vegetables

Preparing vegetables using proper techniques promotes safety and helps ensure that vegetables are appropriate for the dish they are to enhance.

Prepare the following vegetables using proper knife skills. Have an instructor check the result and sign on the lines below when complete.

1. Clean and prepare a leek.

Cleaning and Preparing a Leek Proficiency

Instructor:____ Date:________________

2. Prepare an artichoke.

Preparing an Artichoke Proficiency

Instructor:____ Date:________________

3. Prepare a tomato concassé.

Preparing Tomato Concassé Proficiency

Instructor:____ Date:________________

4. Core a pepper.

Coring a Pepper Proficiency

Instructor:____ Date:________________

5. Fire-roast a pepper.

Fire-Roasting a Pepper Proficiency

Instructor:____ Date:________________

Chapter 14 Review
Potatoes, Grains, and Pastas

Section 14.1 Potatoes

Name ______________________ Date ______________

True-False

T F 1. Potatoes can be purchased fresh, frozen, canned, dehydrated, or as processed items.

T F 2. The flavor of a sweet potato is somewhat dry and starchier than a yam.

T F 3. Potatoes should be clean and have shallow eyes.

T F 4. When storing fresh potatoes, refrigeration is recommended.

T F 5. Waxy potatoes are the preferred type of potato for baking, frying, mashing, puréeing, and casseroles.

Multiple Choice

__________ 1. A russet potato is a ___ potato with thin brown skin, an elongated shape, and shallow eyes.
- A. sweet
- B. new
- C. waxy
- D. mealy

__________ 2. ___ potatoes include red potatoes, yellow potatoes, and fingerling potatoes.
- A. Waxy
- B. Mealy
- C. Yukon
- D. Sweet

__________ 3. The flesh of a ___ potato has a buttery, nutty flavor and has a yellowish color after cooking.
- A. sweet
- B. fingerling
- C. new
- D. yellow

__________ 4. A ___ potato is a small, tapered, waxy potato with butter-colored flesh and tan, red, or purple skin.
- A. russet
- B. fingerling
- C. new
- D. sweet

_______________ **5.** Most graded potatoes are U.S. ___.
A. Fancy
B. Commercial
C. No. 1
D. No. 2

Completion

_______________ **1.** A(n) ___ is a tuber that grows on a vine and has a paper-thin skin and flesh that ranges in color from ivory to dark orange.

_______________ **2.** Potatoes that are sprouting or that have a green-colored flesh or skin should not be eaten because they contain a toxin called ___.

_______________ **3.** A(n) ___ potato refers to any variety of potato that is harvested before the sugar is converted to starch.

_______________ **4.** A(n) ___ potato is a type of potato that is higher in starch and lower in moisture than other types of potatoes.

_______________ **5.** A potato is a round, oval, or elongated ___ that is the only edible part of the potato plant.

Chapter 14 Review
Potatoes, Grains, and Pastas

Section 14.2 Preparing Potatoes

Name ______________________________ **Date** ______________________

True-False

T F **1.** French fries need to be blanched and then immediately deep-fried while they are still hot.

T F **2.** Baked potatoes have a golden color and smoky flavor.

T F **3.** Lightly squeezing a baked potato will indicate whether it is done.

Multiple Choice

______________ **1.** ___ potatoes make the best baked potatoes.
- A. Yukon gold
- B. Purple
- C. Fingerling
- D. Russet

______________ **2.** Potatoes that are ___ or boiled first should be removed from the cooking liquid and allowed to steam dry before they are sautéed.
- A. baked
- B. steamed
- C. roasted
- D. grilled

______________ **3.** If simmered potatoes are to be mashed, ___ potatoes are the best choice.
- A. mealy
- B. waxy
- C. new
- D. red

Completion

______________ **1.** Potatoes help thicken sauces by releasing ___ during the cooking process.

______________ **2.** ___ is the process of topping a dish with a thick sauce, cheese, or bread crumbs and then browning it in a broiler or high-temperature oven.

______________ **3.** A(n) ___ is a baked dish containing a starch, other ingredients (such as meat or vegetables), and a sauce.

Culinary Arts
PRINCIPLES AND APPLICATIONS
STUDY GUIDE

Chapter 14 Review
Potatoes, Grains, and Pastas

Section 14.3 Grains

Name ______________________________ **Date** ______________

True-False

T F **1.** A milled grain is a refined grain with a pearl-like appearance that results from having been scrubbed and tumbled to remove the bran.

T F **2.** Spelt is commonly mistaken for farro because of their similar appearances.

T F **3.** Quinoa is a small, round, gluten-free grain that is classified as a complete protein.

T F **4.** Kasha is roasted buckwheat.

T F **5.** Grains should be stored in a warm, dry place in an airtight container to keep out moisture.

T F **6.** Farro is often added to soups and stews because it is a natural thickener.

T F **7.** A wheat berry is a chewy wheat kernel with only the husk removed.

T F **8.** Semolina is a granular product that results from milling corn.

T F **9.** Rice is a staple food source for two-thirds of the world's population.

T F **10.** Arborio rice is a long-grain rice.

Multiple Choice

______________ **1.** A native of Asia, ___ resembles couscous, yet is prepared like rice.
- A. millet
- B. spelt
- C. buckwheat
- D. farro

______________ **2.** A ___ grain, also known as a rolled grain, is a refined grain that has been rolled to produce a flake.
- A. milled
- B. cracked
- C. whole
- D. flaked

_______________ **3.** ___ oats are oat groats that have been toasted and cut into small pieces.
A. Milled
B. Cracked
C. Steel-cut
D. Old-fashioned

_______________ **4.** A ___ grain is a refined grain that has been ground into a fine meal or powder.
A. flaked
B. milled
C. pearled
D. cracked

_______________ **5.** Basmati rice is a ___-grain rice that only expands lengthwise when it is cooked.
A. short
B. medium
C. long
D. large

_______________ **6.** Rye flour is heavier and ___ in color than wheat flours.
A. darker
B. lighter
C. softer
D. brighter

_______________ **7.** ___ is an ancient grain that is native to Italy.
A. Millet
B. Farro
C. Kasha
D. Triticale

_______________ **8.** The ___ is the tough outer layer of grain that covers the endosperm.
A. germ
B. husk
C. hull
D. bran

_______________ **9.** ___ is a tiny, round pellet made from durum wheat that has had both the bran and germ removed.
A. Semolina
B. Wheat berry
C. Couscous
D. Hominy

_______________ **10.** ___ is coarsely ground corn.
A. Cornmeal
B. Hominy
C. Grits
D. Couscous

Completion

______________ **1.** A refined grain is a grain that has been processed to remove the germ, ___, or both.

______________ **2.** The ___ is the largest component of a grain kernel and is milled to produce flours and other grain products.

______________ **3.** Hominy is the hulled kernels of ___ that have been stripped of their bran and germ and then dried.

______________ **4.** A grain is the edible fruit, in the form of a kernel or seed, of a(n) ___.

______________ **5.** Rice is classified by the ___ of the grain.

______________ **6.** A cracked grain is a whole kernel of grain that has been cracked by being placed between ___.

______________ **7.** Its high protein content and gluten strength make durum the wheat of choice for making ___ dough.

______________ **8.** ___ is a hybrid grain made by crossing rye and wheat.

______________ **9.** A buckwheat ___ is a crushed, coarse piece of whole-grain buckwheat that can be prepared like rice.

______________ **10.** The ___ is the smallest part of a grain kernel.

Matching

____________________ **1.** Brown rice

____________________ **2.** Short-grain rice

____________________ **3.** Basmati rice

____________________ **4.** Medium-grain rice

____________________ **5.** Corn

____________________ **6.** Long-grain rice

____________________ **7.** Whole grain

____________________ **8.** Grits

____________________ **9.** Oat groat

____________________ **10.** Rolled oats

A. Oat grain that only has the husk removed

B. Rice that contains slightly less starch than short-grain rice but is still glossy and slightly sticky when cooked

C. Grain that has only had the husk removed

D. Oats that have been steamed and flattened into small flakes

E. Coarse type of meal made from ground corn or hominy

F. Rice that has had only the husk removed

G. Rice that is long and slender and remains light and fluffy after cooking

H. Long-grain rice that only expands lengthwise when it is cooked

I. Rice that is almost round in shape and has moist grains that stick together when cooked

J. Cereal grain cultivated from an annual grass that bears kernels on large woody cobs called ears

CHAPTER 14 REVIEW
POTATOES, GRAINS, AND PASTAS

SECTION 14.4 PREPARING GRAINS

Name ______________________________ **Date** ________________

True-False

T F **1.** Risotto and pilaf are two grain preparations that are sautéed and then simmered.

T F **2.** With the exception of flaked grains and hominy, grains do not expand in volume when they are cooked.

T F **3.** Sautéing is the most common method of cooking grains.

Multiple Choice

______________ **1.** Cooked grains that are being held for service should be kept at ___°F or above.
- A. 70
- B. 105
- C. 120
- D. 140

______________ **2.** When using the pilaf method, the flavoring ingredients and grains are ___ before liquid is added.
- A. deep-fried
- B. baked
- C. sautéed
- D. steamed

______________ **3.** A classic pilaf is finished ___.
- A. on the stove
- B. in a pressure cooker
- C. in a slow cooker
- D. in the microwave

Completion

______________ **1.** ___ is a classic Italian dish traditionally made with Arborio rice.

______________ **2.** When cooking grains, stocks may be used instead of ___ to add flavor.

______________ **3.** Risotto is cooked slowly to release the ___ from the grain, resulting in a creamy finished product.

Name ______________________________ Date ____________________

True-False

T F **1.** Common types of tube pastas include conchiglie, farfalle, gemelli, orzo, and rotini.

T F **2.** Pasta is a term for rolled or extruded products made from a dough composed of flour, water, salt, oil, and sometimes eggs.

T F **3.** Long-tube pastas are used in soups more often than short-tube pastas.

T F **4.** Rotini, manicotti, and penne are examples of ribbon pasta.

T F **5.** Tortelloni resemble pot stickers, are larger than tortellini, and hold more filling than tortellini.

T F **6.** Pasta can be made fresh or purchased in dried or frozen form.

Multiple Choice

______________ **1.** ___, also known as cannelloni, are large, round tubes of pasta approximately 4 inches long and 1 inch in diameter.
A. Manicotti
B. Rigatoni
C. Penne
D. Ziti

______________ **2.** ___, also known as bow tie pasta, are flat squares of pasta that are pinched in the center in the shape of bow ties.
A. Stelline
B. Gemelli
C. Fusilli
D. Farfalle

______________ **3.** ___ are long, thin, flat strips of pasta approximately ¼ inch wide.
A. Capellini
B. Spaghetti
C. Fettuccine
D. Linguini

______ **4.** ___, also known as cavatappi, are twisted, hollow tubes of pasta with a ribbed surface.
A. Pipettes
B. Cellentani
C. Ditalini
D. Fusilli

______ **5.** ___ pastas include ravioli, tortellini, and tortelloni.
A. Tube
B. Ribbon
C. Shaped
D. Stuffed

______ **6.** Pasta has a ___ flavor that does not compete with the flavors of other ingredients or with the sauces that are often added to it.
A. mild
B. sweet
C. spicy
D. nutty

Completion

______ **1.** When pasta dough is soft, it can be shaped by rolling it flat and cutting it to the desired size or by forcing it through the metal dies of a(n) ___ to create various shapes.

______ **2.** ___ provides the strength to hold the shape, form, and texture of pasta dough when cooked.

______ **3.** The four general types of pasta are shaped, tube, ribbon, and ___.

______ **4.** ___ are small, ridged, bowl-shaped pasta.

______ **5.** Cellentani, ditalini, elbows, manicotti, penne, pipettes, rigatoni, and ziti are common types of ___ pasta.

______ **6.** The shape of the pasta chosen for a given dish is often determined by the ___ and how it will cling to the particular pasta shape.

Matching

______ **1.** Tube pasta

______ **2.** Shaped pasta

______ **3.** Ribbon pasta

______ **4.** Stuffed pasta

A. Pasta that has been extruded into a complex shape such as a corkscrew, bowtie, shell, flower, or star

B. Pasta that has been pushed through an extruder and then fed through a cutter that cuts the tubes to the desired length

C. Pasta that has been formed by hand or machine to hold fillings

D. Thin, round strand or flat, ribbonlike strand of pasta

Chapter 14 Review
Potatoes, Grains, and Pastas

Section 14.6 Preparing Pasta

Name ______________________________ **Date** ____________________

True-False

T F **1.** When making tortellini, thin pasta dough is cut into squares and a filling is placed in the center of each square.

T F **2.** If pasta is made fresh, the dough should never be kneaded.

T F **3.** Parcooked pasta can be rinsed in warm water to bring it to the temperature needed for service.

T F **4.** When pasta is cooked al dente, there should be a slight resistance in the center of the pasta when it is chewed.

T F **5.** Squid ink can provide a salty, metallic flavor, but it will not color the pasta.

Multiple Choice

______________ **1.** When preparing ravioli, a pastry brush is used to ___ wash the pasta around each portion of filling.
- A. water
- B. milk
- C. oil
- D. egg

______________ **2.** When making fresh pasta dough, the dough should be covered and rested for ___ minutes before rolling the dough to the appropriate thickness.
- A. 5
- B. 12
- C. 20
- D. 30

______________ **3.** A pound of fresh pasta yields approximately ___ lb of cooked pasta.
- A. ¼–½
- B. 1–1½
- C. 2–2½
- D. 3–3½

______________ **4.** Dried spaghetti cooks in approximately ___ minutes.
A. 2–3
B. 4–5
C. 7–8
D. 10–12

______________ **5.** Fresh pasta cooks in approximately ___ minutes.
A. 1–2
B. 3–5
C. 7–9
D. 10–12

Completion

______________ **1.** To cook pasta, one gallon of water per ___ of pasta is placed in a stockpot or steam-jacketed kettle.

______________ **2.** Pasta should be cooked ___, meaning "to the tooth."

______________ **3.** To make ___, one sheet of pasta is topped with a small amount of filling, and a second sheet of pasta is laid on top.

______________ **4.** Pasta at least ___ in volume when cooked.

Chapter 14 Review
Potatoes, Grains, and Pastas

Section 14.7 Asian Noodles

Name ______________________________ **Date** ______________

True-False

T F **1.** Lo mein noodles have no calories, carbohydrates, gluten, or fat.

T F **2.** Asian noodles are generally not boiled like pasta.

T F **3.** Buckwheat noodles, also known as soba noodles, are brown-gray noodles made from buckwheat flour.

Multiple Choice

______________ **1.** ___ noodles that have been deep-fried and then dehydrated are sold as ramen noodles.
- A. Somen
- B. Shirataki
- C. La mian
- D. Lo mein

______________ **2.** Cellophane noodles and shirataki noodles are two common types of ___ noodles.
- A. vegetable
- B. egg
- C. rice
- D. buckwheat

______________ **3.** ___ noodles are yellow egg noodles that resemble spaghetti.
- A. Cellophane
- B. Hokkien
- C. Lo mein
- D. Udon

Completion

______________ **1.** Udon, somen, and la mian noodles are all made from ___.

______________ **2.** ___ noodles, also known as glass noodles, are made from mung bean starch.

______________ **3.** ___ noodles, also known as Cantonese noodles, are egg noodles in varying widths.

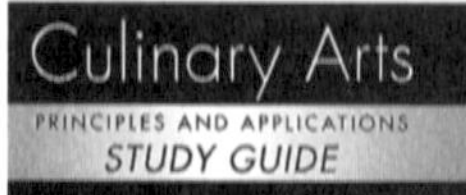

Name ______________________________ **Date** ________________

Activity: Preparing Potatoes

There are many ways to prepare potatoes. Different varieties of potatoes can produce different results in texture and flavor, depending on the cooking method used.

Wash and slice a mealy potato and a waxy potato into ¼-inch-thick slices. Blanch and then sauté the potatoes separately until they are done.

1. Describe the texture and flavor of the mealy potato as prepared.

2. Describe the texture and flavor of the waxy potato as prepared.

3. Which variety of potato produces the best result with this cooking method?

Wash and slice a mealy potato and a waxy potato into ¼-inch-thick slices. Sauté the potatoes separately without blanching.

4. Describe the texture and flavor of the mealy potato as prepared.

5. Describe the texture and flavor of the waxy potato as prepared.

6. Which variety of potato produces the best result with this cooking method?

7. Which method produces the best overall result: blanching and sautéing or sautéing without blanching?

Peel four mealy potatoes and simmer them until soft. Mash the potatoes in a mixer or by hand, and add ¼ cup milk and ¼ cup butter. Season the potatoes with salt and pepper to taste.

8. Describe the texture and flavor of the mashed potatoes as prepared.

Peel four waxy potatoes and simmer until soft. Mash the potatoes in a mixer or by hand, and add ¼ cup milk and ¼ cup butter. Season the potatoes with salt and pepper to taste.

9. Describe the texture and flavor of the mashed potatoes as prepared.

10. Which variety of potato produces the better mashed potato: mealy potatoes or waxy potatoes?

Wash a mealy potato and a waxy potato and simmer each potato until it is soft.

11. Describe the texture and flavor of the mealy potato as prepared.

12. Describe the texture and flavor of the waxy potato as prepared.

13. Which variety of potato produces the best result with this cooking method?

Prepare a pan of rissole (oven-browned) potatoes using mealy potatoes. Prepare a second pan of rissole using waxy potatoes.

14. Describe the texture and flavor of the mealy potatoes as prepared.

15. Describe the texture and flavor of the waxy potatoes as prepared.

16. Which variety of potato produces the best result with this cooking method?

Prepare French-fried potatoes using both mealy potatoes and waxy potatoes.

17. Describe the texture and flavor of the mealy potatoes as prepared.

18. Describe the texture and flavor of the waxy potatoes as prepared.

19. Which variety of potato produces the best result with this cooking method?

Activity: Analyzing Cooking Yields of Grains

Simmering is the most common method of cooking grains. The grain is simply added to a measured amount of boiling cooking liquid. The appropriate amount of cooking liquid depends on the type and amount of grain being cooked.

Prepare brown basmati rice using 1 cup dry rice and 2 cups liquid.

1. What is the approximate yield of the cooked brown basmati rice?

Prepare couscous using 1 cup dry couscous and 1.25 cup liquid.

2. What is the approximate yield of the cooked couscous?

Prepare quinoa using 1 cup dry quinoa and 2 cups liquid.

3. What is the approximate yield of the cooked quinoa?

4. What conclusion can be drawn about cooking yields for grains?

Activity: Identifying Pasta Shapes

Pasta can be formed into many different shapes and sizes. The shape of the pasta chosen for a dish is often determined by the sauce and how it will cling to the particular pasta shape.

Match each type of pasta to its image.

_______________ **1.** Rigatoni

_______________ **2.** Cellentani

_______________ **3.** Tortellini

_______________ **4.** Egg noodles

_______________ **5.** Fiori

_______________ **6.** Orzo

_______________ **7.** Linguine

_______________ **8.** Penne

_______________ **9.** Tortelloni

_______________ **10.** Elbows

_______________ **11.** Fusilli

_______________ **12.** Orecchiette

_______________ **13.** Lasagna

_______________ **14.** Fettuccine

_______________ **15.** Farfalle

Barilla America, Inc.
Ⓐ

Barilla America, Inc.
Ⓑ

Barilla America, Inc.
Ⓒ

Barilla America, Inc.
Ⓓ

Barilla America, Inc.
Ⓔ

Barilla America, Inc.
Ⓕ

Barilla America, Inc.
Ⓖ

Barilla America, Inc.
Ⓗ

Barilla America, Inc.
Ⓘ

Barilla America, Inc.
Ⓙ

Barilla America, Inc.
Ⓚ

Barilla America, Inc.
Ⓛ

Barilla America, Inc.
Ⓜ

Barilla America, Inc.
Ⓝ

Barilla America, Inc.
Ⓞ

Activity: Comparing House-Made and Convenience Pastas

Pasta is often purchased ready-to-use because it is convenient, inexpensive, available in many different varieties, and less labor intensive when compared to pasta made from scratch. However, some chefs prefer to make pasta from scratch because it allows the chef to flavor, color, and fill the pasta as desired.

Prepare a house-made pasta dough. Roll and cut linguine noodles with half of the prepared dough. Roll and form ravioli using a prepared ricotta cheese filling with the other half of the prepared dough. Cook the pasta al dente and taste it. Then, respond to the following items.

1. Describe the texture and flavor of the house-made linguine.

2. How much time did it take to prepare the house-made linguine from start to finish?

3. Estimate the cost of ingredients used in the house-made linguine.

4. Describe the texture and flavor of the house-made ravioli.

5. How much time did it take to prepare the house-made ravioli from start to finish?

6. Estimate the cost of ingredients used in the house-made ravioli.

Obtain a convenience (ready-to-use) linguine and cheese ravioli. Cook the pasta al dente and taste it. Then, respond to the following items.

7. Describe the texture and flavor of the convenience linguine.

8. How much time did it take to prepare the convenience linguine from start to finish?

9. Estimate the cost of the convenience linguine.

10. Describe the texture and flavor of the convenience ravioli.

11. How much time did it take to prepare the convenience ravioli from start to finish?

12. Estimate the cost of the convenience ravioli.

13. Compare and contrast the texture and flavor of the house-made pastas with the convenience pastas.

Activity: Researching Potatoes, Grains, or Pastas

Potatoes, grains, and pastas have a rich and diverse history. They have been prepared and served throughout the ages using various cooking methods and presentations that highlight their versatility.

Choose a type of potato, grain, or pasta and write an essay using the following outline. Include five recipes that use the chosen topic as a main ingredient. The recipes should use different methods of preparation and serve as different parts of the menu such as appetizers, soups, salads, entrées, and desserts.

I. Introduction
II. History of the food
  A. Country of origin
  B. Economic influences
III. Contemporary use
  A. Countries where the food is a staple item
  B. Cultural influences
  C. Common preparations
IV. Recipes
V. Conclusion

1. Prepare and present a report about the chosen potato, grain, or pasta using the outline provided. Use visuals and demonstrations to enhance the report. Include a list of the sources used for the project.

Name ______________________________ Date ____________________

True-False

T F **1.** Butterhead lettuces have a curly, edible leaf with a slightly bitter flavor.

T F **2.** Arugula has a mild peppery flavor and is most often served by itself.

T F **3.** Looseleaf lettuces have rich-colored, soft leaves and a very mild flavor.

T F **4.** Dandelion greens can taste quite sweet.

T F **5.** Radicchio is a small, compact head of red leaves, similar to a small head of red cabbage.

Multiple Choice

______________ **1.** ___ has a slender, tightly packed, elongated head that forms a point.
A. Curly endive
B. Belgian endive
C. Frisée
D. Radicchio

______________ **2.** Boston lettuce and ___ lettuce are two varieties of butterhead lettuce.
A. iceberg
B. gem
C. romaine
D. Bibb

______________ **3.** A(n) ___ salad is a sweet salad usually consisting of nuts, fruits, and sweeter vegetables, such as carrots.
A. dessert
B. side
C. appetizer
D. separate course

______________ **4.** ___ is a large, dark-green, edible leaf with white or reddish stalks.
A. Arugula
B. Chard
C. Kale
D. Spinach

______________ **5.** A ___ is an edible strand with an attached bud that comes from a germinated bean or seed.

A. stalk
B. flowering herb
C. sprout
D. seed

Completion

______________ **1.** A(n) ___ salad is a small salad served to accompany a main course.

______________ **2.** ___ are the first sprouting leaves of an edible plant.

______________ **3.** ___ greens are a mix of young greens that range in color, texture, and flavor.

______________ **4.** ___ cabbage is a conical-shaped head of tender, crinkly, edible leaves that are blue-green on the exterior and pale green on the interior.

______________ **5.** A(n) ___ is also known as a pepita.

Chapter 15 Review
Garde Manger Fundamentals

Section 15.2 Preparing Salad Greens

Name ______________________ Date ______________

True-False

T F 1. All salad greens should be stored between 35°F and 38°F, which is colder than the storage temperature of other produce.

T F 2. The ribs of romaine lettuce must be removed prior to preparation.

T F 3. If salad greens look clean, it is not essential to wash them before use.

T F 4. If salad greens are handled incorrectly, they can bruise, brown, and lose crispness.

T F 5. Salad greens grow high above the ground and are typically clean.

Multiple Choice

__________ 1. The first step when washing salad greens is to ___.
- A. rinse the leaves under hot water
- B. cut or tear the greens to the desired size
- C. submerge the greens in soapy water
- D. wipe away dirt with a chemical sanitizer

__________ 2. Loose greens, such as chicory and kale, often ___ in the center rib.
- A. collect dirt
- B. wilt
- C. loose flavor
- D. discolor

__________ 3. Salad greens must be stored away from fruits that emit ___.
- A. carbon dioxide
- B. nitrogen
- C. ethylene gas
- D. hydrogen

Completion

__________ 1. Using a(n) ___ is the best way to dry lettuce.

__________ 2. Delicate greens, such as butterhead lettuces, deteriorate ___ than firmer varieties such as romaine.

____________________ **3.** The flavorful ___ are not removed from romaine lettuce.

____________________ **4.** Solidly packed head lettuces, such as iceberg, must have the ___ removed and be cut to an appropriate size before washing the leaves.

____________________ **5.** All salad greens must be ___ completely after washing or they will become limp in a short amount of time.

Chapter 15 Review
Garde Manger Fundamentals

Section 15.3 Salad Ingredients and Dressings

Name ______________________________ **Date** ____________________

True-False

T F **1.** Grains used in a salad should be cooked slightly less than al dente.

T F **2.** Only raw vegetables are used in salads.

T F **3.** When preparing mayonnaise, the vinegar is added very slowly until the emulsion begins to form and the mixture thickens.

Multiple Choice

______________ **1.** A(n) ___ is a combination of two unlike liquids that have been forced to bond with each other.

A. emulsion
B. suspension
C. dressing
D. dip

______________ **2.** The ingredients used to prepare an emulsified vinaigrette must be ___ before starting the preparation.

A. frozen
B. cold
C. room temperature
D. hot

______________ **3.** A ___ is an edible root, bulb, tuber, stem, leaf, flower, or seed of a nonwoody plant.

A. protein
B. vegetable
C. starch
D. fruit

Completion

______________ **1.** In a basic French vinaigrette, the ratio of oil to vinegar is ___.

______________ **2.** Rapidly whisking oil and vinegar together creates a(n) ___.

______________ **3.** Mayonnaise is an example of a(n) ___ emulsion.

Name ______________________________ **Date** ________________

True-False

T F **1.** For best results, gelatin should be placed in the freezer.

T F **2.** Popular composed salads include egg salad, chicken salad, and tuna salad.

T F **3.** When making vegetable salads, knowing how the vegetables react to acidity is important.

Multiple Choice

______________ **1.** To prevent rapid discoloration, some fruits in fruit salads should be dipped or lightly tossed in liquids that contain ___.

A. vanilla
B. gelatin
C. oil
D. citric acid

______________ **2.** The ___ of a composed salad consists of the main ingredients.

A. base
B. body
C. garnish
D. dressing

______________ **3.** Fruit-flavored gelatin powder packaged for commercial use typically contains 1 lb 8 oz of gelatin powder and requires a total of ___ gal. of water or fruit juice to be added during preparation.

A. ¼
B. ½
C. 1
D. 1½

Completion

______________ **1.** A(n) ___ salad is a mixture of leafy greens, such as lettuce, spinach, chicory, or fresh herbs, and other ingredients, such as fruits, vegetables, nuts, cheese, meats, and croutons, served with a dressing.

_______________ **2.** The ___ of a composed salad is typically a bed of salad greens, fruits, or vegetables.

_______________ **3.** In a(n) ___ salad, mayonnaise, vinaigrettes, yogurt, and creamy dressings act as a binding agent.

Matching

_______________ **1.** Composed salad

_______________ **2.** Gelatin salad

_______________ **3.** Bound salad

_______________ **4.** Vegetable salad

_______________ **5.** Fruit salad

A. Salad made from flavored gelatin

B. Salad made by combining a main ingredient, often a protein, with a binding agent such as mayonnaise or yogurt and other flavoring ingredients

C. Salad that is primarily made of fruits

D. Salad that consists of a base, body, garnish, and dressing attractively arranged on a plate

E. Salad that is primarily made of vegetables

Name ____ Date ____________________

True-False

T F **1.** Brie is a creamy, white soft cheese with a strong odor and a sharp taste.

T F **2.** Roquefort is a blue-veined cheese made from sheep milk and is characterized by a sharp, tangy flavor.

T F **3.** Hard cheese varieties do not grate well.

T F **4.** The blue vein that runs through certain cheeses indicates where the needle was inserted and the mold spores were released.

T F **5.** Gruyère is a semisoft cheese of Italian origin that has a light-brown rind, small holes, and a nutty flavor.

Multiple Choice

______________ **1.** Common examples of dry-rind semisoft cheeses are Monterey Jack, Bel Paese®, and ___.
A. Edam
B. Limburger
C. Camembert
D. Havarti

______________ **2.** A ___ cheese is a cheese that is not aged or allowed to ripen.
A. hard
B. semisoft
C. soft
D. fresh

______________ **3.** Edam is a ___-rind semisoft cheese made from cow milk and has a firm, crumbly texture.
A. washed
B. waxed
C. dry
D. wet

______________ **4.** Cheese is most often made from milk that has been coagulated or curdled using ___.
A. rennet
B. whey
C. mold
D. brine

______ **5.** ___ is a cream cheese of Italian origin that has a smooth texture, is white or pale yellow in color, and has a buttery, somewhat sweet flavor.
A. Ricotta
B. Manchego
C. Mascarpone
D. Neufchâtel

Completion

______ **1.** The ___ is the thick, casein-rich part of coagulated milk.

______ **2.** Common examples of ___-rind semisoft cheeses are brick, Limburger, Muenster, and Port Salut.

______ **3.** Asiago is a(n) ___ cheese with a nutty, toastlike flavor.

______ **4.** Chèvre is a fresh cheese made from ___ milk.

______ **5.** A(n) ___ is a processed food made of natural cheeses that may include additional ingredients such as emulsifiers.

Matching

______ **1.** Hard cheese

______ **2.** Semisoft cheese

______ **3.** Fresh cheese

______ **4.** Waxed-rind cheese

______ **5.** Blue-veined cheese

______ **6.** Dry-rind cheese

______ **7.** Washed-rind cheese

______ **8.** Soft cheese

______ **9.** Grating cheese

______ **10.** Cold-pack cheese

______ **11.** Processed cheese

______ **12.** Processed cheese food

A. Hard, crumbly, dry cheese grated or shaved onto food prior to service

B. Semisoft cheese produced by dipping a wheel of freshly made cheese into a liquid wax and allowing the wax to harden

C. Cheese-based product that may contain as little as 51% cheese

D. Semisoft cheese with an exterior rind that is washed with a brine, wine, olive oil, nut oil, or fruit juice

E. Cheese that is not aged or allowed to ripen

F. Creamy cheese product made by blending natural cheeses without the addition of heat

G. Semisoft cheese that is allowed to ripen through exposure to air

H. Cheese produced by inserting harmless live mold spores into the center of ripening cheese with a needle

I. Firm, somewhat pliable and supple cheese with a slightly dry texture and buttery flavor

J. Cheese that is firmer than a soft cheese but not as hard as a hard cheese

K. Cheese that has been sprayed with a harmless live mold to produce a thin skin or rind

L. Blend of fresh and aged natural cheeses that are heated and melted together

Name ______________________________ Date ____________________

True-False

T F **1.** When refrigerated properly, fresh cheeses keep for about one month.

T F **2.** Cheese should ideally be served at room temperature.

T F **3.** Cheese becomes tough and stringy when overheated.

T F **4.** When refrigerated properly, soft cheeses keep for several months.

Multiple Choice

____________ **1.** When refrigerated properly, semisoft cheeses keep for ___.
A. three to six days
B. one week
C. two to three weeks
D. several months

____________ **2.** If cheese has been refrigerated, it should be removed approximately ___ prior to service for best flavor and texture.
A. 5 minutes
B. 15 minutes
C. 40 minutes
D. 1 hour

____________ **3.** Cheeses are generally cooked at the lowest temperature possible to prevent proteins in the cheese from ___.
A. hardening
B. softening
C. converting to starches
D. converting to fats

____________ **4.** In ___, cheese is traditionally served with a continental breakfast.
A. Asia
B. Europe
C. South America
D. Central America

Completion

______________ **1.** Cheese should be added at the ___ of the cooking process so it melts smoothly and incorporates evenly.

______________ **2.** ___ is a dairy product that contains a considerable amount of butterfat, absorbs odors easily, spoils relatively quickly, and can dry out if left exposed to air.

______________ **3.** When preparing a cheese platter, it is important to include an assortment of cheeses with various ___, flavors, and degrees of ripeness.

Chapter 15 Review
Garde Manger Fundamentals

Section 15.7 Hors d'Oeuvres and Appetizers

Name ______________________________ **Date** ____________________

True-False

T F **1.** Canapés are composed of a base, a spread, and an edible garnish.

T F **2.** Tartare presentations often include colorful selections of meats, cheeses, and marinated, pickled, or grilled vegetables.

T F **3.** A brochette is a food that is speared onto a wooden, metal, or natural skewer and then grilled or broiled.

T F **4.** The batter and breading used in casual dining restaurants is often heavier than the type used in upscale dining environments.

T F **5.** To prepare skewered starters, the wooden skewers are soaked in oil for at least 30 minutes prior to use.

T F **6.** To prepare canapés, the bread base is sliced lengthwise into evenly sliced pieces.

Multiple Choice

______________ **1.** A raw bar is a presentation of a variety of raw and steamed ___ presented and served on a bed of ice.
- A. fruits
- B. vegetables
- C. seafood
- D. meat

______________ **2.** ___ is served from carts that are stacked with round steamer baskets and contain items such as pot stickers and wontons.
- A. Dim sum
- B. Tapas
- C. Antipasti
- D. Charcuterie

______________ **3.** Makizushi is a type of sushi made with vinegar-seasoned cooked rice layered with other ingredients and rolled in a dried seaweed paper called ___.
- A. tempeh
- B. miso
- C. kimchi
- D. nori

______________ **4.** Cold smoking exposes food to smoke at temperatures below ___.
A. 100°F
B. 125°F
C. 150°F
D. 175°F

______________ **5.** A(n) ___ is food that is larger than a single bite and is typically served as the first course of a meal.
A. hors d'oeuvre
B. appetizer
C. amuse-bouche
D. canapé

______________ **6.** A(n) ___ is a miniature, boat-shaped pastry shell that contains a savory or sweet filling.
A. profiterole
B. phyllo shell
C. barquette
D. brochette

Completion

______________ **1.** A(n) ___ is a puff pastry that is filled with a savory filling.

______________ **2.** A(n) ___ is a single, bite-sized portion of food selected by the chef and not ordered by the guest.

______________ **3.** Grilling and roasting vegetables causes the natural ___ to caramelize and intensify in flavor.

______________ **4.** ___ is thin slices of meats or seafood that are served raw.

______________ **5.** A crudité is a group of raw ___ arranged on a platter and served with a dipping sauce.

______________ **6.** A(n) ___ is an elegant, bite-size portion of food that is creatively presented and served apart from a meal.

Chapter 15 Review
Garde Manger Fundamentals

Section 15.8 Charcuterie

Name ______________________________ **Date** ________________

True-False

T F **1.** When grinding forcemeats, all of the ingredients and equipment must be room temperature.

T F **2.** A hard fat, such as pork fatback, is the least desirable fat to use in a forcemeat.

T F **3.** Pork is usually added to forcemeats without a pork base in a ratio of 1 part pork to 2 parts other protein.

Multiple Choice

________________ **1.** A ___ forcemeat is a forcemeat made from less-fibrous proteins such as poultry, seafood, veal, or pork.
A. gratin
B. straight
C. mousseline-style
D. country-style

________________ **2.** ___ is often added to forcemeats as a binding agent and as a flavoring.
A. Olive oil
B. Curing salt
C. Mayonnaise
D. Liver

________________ **3.** ___ is the most important seasoning to add to a forcemeat.
A. Salt
B. Pepper
C. Vinegar
D. Citric acid

Completion

________________ **1.** ___ is the art of making sausages and other preserved items such as pâtés, terrines, galantines, and ballotines.

________________ **2.** A(n) ___ is a mixture of raw or cooked meat, poultry, seafood, or vegetables, fat, seasonings, and sometimes other ingredients.

________________ **3.** Pâtés and terrines are typically prepared using a(n) ___ forcemeat.

Matching

__________ **1.** Gratin forcemeat

__________ **2.** Straight forcemeat

__________ **3.** Forcemeat

__________ **4.** Country-style forcemeat

__________ **5.** Curing salt

__________ **6.** Emulsified forcemeat

A. Combination of table salt and sodium nitrate

B. Coarsely ground forcemeat with a strong flavor

C. Forcemeat that consists of seasoned ground meat emulsified with ground fat

D. Forcemeat made from a ratio of five parts protein, to four parts fat, to three parts crushed ice

E. Forcemeat made from meat that is partially seared prior to grinding and a starchy binding agent

F. Mixture of raw or cooked meat, poultry, seafood, or vegetables, fat, seasonings, and sometimes other ingredients

SECTION 15.9 PREPARING AND PLATING CHARCUTERIE

Name ______________________________ **Date** ______________

True-False

T F **1.** Chaud froid is a savory jelly made from a consommé or clarified stock that produces a clear finish when coating foods.

T F **2.** A terrine is a forcemeat that is baked in a mold without a crust.

T F **3.** Because charcuterie products have a high ratio of fat added while grinding, keeping items warm results in a better emulsification.

T F **4.** Galantines are always served cold, whereas ballotines may be served hot or cold.

T F **5.** The common ratio of fat to lean meat in most sausages is 1 to 3 or about 33% fat.

Multiple Choice

______________ **1.** ___ casings are the most tender natural casings and are no more than 1 inch in diameter.
A. Deer
B. Cattle
C. Hog
D. Sheep

______________ **2.** Like aspic, ___ protects foods from drying out when exposed to air.
A. chaud froid
B. a glaze
C. a terrine
D. salt

______________ **3.** Charcuterie items are commonly presented ___.
A. on napkins
B. on spoons
C. on platters
D. in ramekins

______________ **4.** Smoked meats, pâtés, and terrines are often served with gherkins and assorted ___.
A. salsas
B. mustards
C. pestos
D. salts

_______________ **5.** The most common shapes of pâté molds are oval and ___.
A. triangular
B. square
C. rectangular
D. round

Completion

_______________ **1.** The three basic categories of sausage are fresh, cured, and ___.

_______________ **2.** A traditional ___ is made from boned-out poultry stuffed with poultry forcemeat and is chilled, glazed with aspic, and garnished with items included in the filling.

_______________ **3.** Preparing a(n) ___ can be as simple as adding gelatin to heavy cream and a white stock.

_______________ **4.** Aspic is used as a protective coating and as a(n) ___ agent.

_______________ **5.** ___ casings are not edible and must be peeled off prior to eating.

Matching

_______________ **1.** Quenelle

_______________ **2.** Smoked sausage

_______________ **3.** Pâté

_______________ **4.** Natural casing

_______________ **5.** Cured sausage

_______________ **6.** Galantine

_______________ **7.** Ballotine

_______________ **8.** Fresh sausage

_______________ **9.** Collagen casing

_______________ **10.** Synthetic fibrous casing

_______________ **11.** Sausage stuffer

A. Casing produced from a plastic-like synthetic material

B. Type of cured sausage that is smoked during the curing process

C. Small dumpling made from a forcemeat by using two spoons

D. Casing produced from collagen

E. Manual or electric piece of equipment that uses a piston to pump forcemeat through a nozzle into casings

F. Type of raw sausage that contains sodium nitrite mixed into the forcemeat

G. Forcemeat that is typically layered with garnishes and baked in a mold

H. Casing produced from the intestines of sheep, hogs, or cattle

I. Type of sausage that is freshly made and has not been cured, smoked, dried, or further processed in any way

J. Boned poultry leg that has been stuffed with forcemeat

K. Forcemeat that is wrapped in the skin of the animal the meat was taken from, shaped into a cylinder, tied, and poached in a flavorful stock

Name ______________________________ **Date** ____________________

Activity: Identifying Salad Greens

A chef must be able to identify types of salad greens by appearance.

Match each salad green to its image.

_______________ **1.** Red and green Royal Oak

_______________ **2.** Belgian endive

_______________ **3.** Iceberg

_______________ **4.** Frisée

_______________ **5.** Radicchio

_______________ **6.** Arugula

_______________ **7.** Boston

_______________ **8.** Dandelion greens

_______________ **9.** Romaine

_______________ **10.** Kale

_______________ **11.** Spinach

_______________ **12.** Napa cabbage

Tanimura & Antle®
Ⓐ

Tanimura & Antle®
Ⓑ

Frieda's Specialty Produce
Ⓒ

Tanimura & Antle®
Ⓓ

Tanimura & Antle®
Ⓔ

Frieda's Specialty Produce
Ⓕ

Tanimura & Antle®
Ⓖ

Tanimura & Antle®
Ⓗ

Melissa's Produce
Ⓘ

Melissa's Produce
Ⓙ

Tanimura & Antle®
Ⓚ

Tanimura & Antle®
Ⓛ

Activity: Identifying Flavors of Salad Greens

Understanding the different flavors that greens possess can help create unique and appetizing salads.

Classify the following salad greens as sweet, mild, bitter, or peppery. Letters can be used more than once.

______ **1.** Mâche

______ **2.** Belgian endive

______ **3.** Romaine

______ **4.** Red leaf lettuce

______ **5.** Arugula

______ **6.** Radicchio

______ **7.** Spinach

______ **8.** Watercress

______ **9.** Iceberg

______ **10.** Butterhead

______ **11.** Dandelion greens

A. Sweet

B. Mild

C. Bitter

D. Peppery

Activity: Analyzing Salad Dressings

The flavor of a dressing is often the first flavor tasted when eating a salad and should complement the greens and additional ingredients.

Answer the following questions about salad dressings.

1. What is the ratio of oil to vinegar in a vinaigrette?

2. What is the definition of an emulsion?

3. Describe the difference between a temporary and a permanent emulsion.

4. Identify the permanent emulsion from the recipes in chapter 15.

5. Describe the flavor of the permanent emulsion.

6. Identify the temporary emulsion from the recipes in chapter 15.

7. Describe the flavor of the temporary emulsion.

8. What are some ingredients that can be used to flavor a vinaigrette?

9. What are some ingredients that can be used to make a flavored oil?

10. What are some ingredients that can be used to make a flavored vinegar?

Activity: Comparing House-Made and Convenience Salad Dressings

House-made dressings allow the chef control over ingredients and, ultimately, the flavor of the finished product. However, ready-to-use convenience salad dressings can be used in foodservice operations as an alternative to making dressings from scratch.

Prepare house-made buttermilk ranch dressing and a convenience buttermilk ranch dressing. Respond to the following.

1. Identify the ingredients in each dressing.

House-made:

Convenience:

2. Based on equivalent amounts, how much does each dressing cost?

House-made:

Convenience:

3. Based on equivalent serving sizes, compare the nutritional information of each dressing.

House-made:

Convenience:

4. Taste each dressing and describe the flavor.

House-made:

Convenience:

5. Rate each dressing on a scale of 1 to 5, with 5 providing the best flavor and quality.

House-made:

Convenience:

6. Describe the advantages of using a house-made dressing.

Activity: Analyzing Salad Types

Salads are classified based on how the salad is plated or served. The six types of salads are tossed, composed, bound, vegetable, fruit, and gelatin.

Respond to the following.

1. Describe a tossed salad.

2. Explain the five steps to follow when preparing a tossed salad.

3. Describe a composed salad.

4. Give an example of bases used for composed salads.

5. Give an example of an item that can serve as the body of a composed salad.

6. Give an example of garnishes used for composed salads.

7. Describe a bound salad.

8. Give an example of a bound salad.

9. Describe a vegetable salad.

10. Give an example of a vegetable salad.

11. Describe a fruit salad.

12. Give examples of fruits that can discolor when exposed to air.

13. Explain how to accelerate the time it takes for gelatin to set.

Activity: Analyzing Cheeses

The flavor and texture of each cheese differs depending on the animal that produced the milk, the diet of that animal, the percentage of butterfat in the cheese, and how long the cheese was aged. Cheeses can be classified as hard, soft, semisoft, fresh, blue-veined, or grating.

Respond to the following. Use purchased cheeses for a reference, if available.

1. Name a hard cheese.

2. Describe the texture and flavor of a hard cheese.

3. Identify a common use for hard cheese.

4. Name a soft cheese.

5. Describe the texture and flavor of a soft cheese.

6. Identify a common use for soft cheese.

7. Name a semisoft cheese.

8. Describe the texture and flavor of a semisoft cheese.

9. Identify a common use for semisoft cheese.

10. Name a fresh cheese.

11. Describe the texture and flavor of a fresh cheese.

12. Identify a common use for fresh cheese.

13. Name a blue-veined cheese.

14. Describe the texture and flavor of a blue-veined cheese.

15. Identify a common use for blue-veined cheese.

16. How are the blue veins in blue-veined cheeses created?

17. Name a hard grating cheese.

18. Describe the texture and flavor of a grating cheese.

19. Identify a common use for hard grating cheese.

Activity: Researching Small Plates of Food

The practice of serving small plates of food as either a starter course or a complete meal is common all over the world. Spain calls their small portions tapas, China has dim sum, Italy has antipasti, and Greece has mezes.

Research the practice of serving small plates of food as either a starter course or a complete meal and write an essay using the following outline.

I. Introduction
II. History of the small plate
  A. Country of origin
  B. Chef or person attributed to creation of the small plate (if any)
  C. Economic influences
III. Contemporary use
  A. Countries where the small plate is commonly served
  B. Cultural influences
  C. Common variations
IV. Recipes
V. Conclusion

1. Prepare and present a report using the outline provided. Use visuals and demonstrations to enhance the report. Include a list of sources used for the research project.

Activity: Scaling Hors d'Oeuvre Recipes

When planning a large event, such as a reception, standardized recipes are often scaled to produce different yields.

Plan an hors d'oeuvres reception for 100 people. Convert the following hors d'oeuvres recipes to serve 100 people using the following formula:

$SF = DY \div OY$

where

SF = scaling factor

DY = desired yield

OY = original yield

1. What is the scaling factor needed to scale scallop rumaki from a total yield of 4 servings to a total yield of 100 servings?

2. Use the scaling factor found in the previous question to scale scallop rumaki from a total yield of 4 servings to a total yield of 100 servings.

Scallop Rumaki (Yield: 100 servings)		
Original Yield ×	**SF** =	**Desired Yield**
4 slices bacon		
8 sea scallops		
salt to taste	—	—
pepper to taste	—	—

3. What is the scaling factor needed to scale Buffalo wings from a total yield of 4 servings to a total yield of 100 servings?

4. Use the scaling factor found in the previous question to scale Buffalo wings from a total yield of 4 servings to a total yield of 100 servings.

Buffalo Wings (Yield: 100 servings)		
Original Yield ×	**SF** =	**Desired Yield**
12 chicken wings		
1¾ c all-purpose flour		
2 tsp salt		
1 tsp pepper		
2 eggs		
½ c milk		
2½ c bread crumbs		
4 tbsp Tabasco® sauce		
1 tbsp vinegar		
1 tsp cayenne pepper		
2 tbsp margarine		

5. What is the scaling factor needed to scale crab Rangoon from a total yield of 8 servings to a total yield of 100 servings?

6. Use the scaling factor found in the previous question to scale crab Rangoon from a total yield of 8 servings to a total yield of 100 servings.

Crab Rangoon (Yield: 100 servings)		
Original Yield ×	SF =	Desired Yield
1 egg		
½ c water		
1 tsp salt		
2 c all-purpose flour		
4 oz cream cheese		
4 oz crab meat, cooked		
½ clove garlic, minced		
1 shallot, finely chopped		
black pepper to taste	—	—

7. What is the scaling factor needed to scale seafood canapés from a total yield of 16 pieces to a total yield of 100 pieces?

8. Use the scaling factor found in the previous question to scale seafood canapés from a total yield of 16 pieces to a total yield of 100 pieces.

Seafood Canapés (Yield: 100 pieces)		
Original Yield ×	SF =	Desired Yield
4 slices white bread		
⅓ c crab meat, flaked		
⅓ c medium or tiny shrimp		
2 tbsp celery, finely chopped		
1 tbsp green onion, finely chopped		
¼ c mayonnaise		
1 tsp sea salt		
2 sprigs fresh dill		

9. What is the scaling factor needed to scale cucumber and smoked salmon canapés with chive cream cheese spread from a total yield of 32 pieces to a total yield of 100 pieces?

10. Use the scaling factor found in the previous question to scale cucumber and smoked salmon canapés with chive cream cheese spread from a total yield of 32 pieces to a total yield of 100 pieces.

Cucumber and Smoked Salmon Canapés with Chive Cream Cheese Spread (Yield: 100 pieces)		
Original Yield ×	**SF** =	**Desired Yield**
1 lb smoked salmon		
1 medium cucumber		
½ c sour cream		
8 oz cream cheese, softened		
1 tbsp chives, finely chopped		
¾ pt grape tomatoes		
1 sprig parsley		

11. What is the scaling factor needed to scale mushroom barquettes from a total yield of 12 pieces to a total yield of 100 pieces?

12. Use the scaling factor found in the previous question to scale mushroom barquettes from a total yield of 12 pieces to a total yield of 100 pieces.

Mushroom Barquettes (Yield: 100 pieces)		
Original Yield ×	**SF** =	**Desired Yield**
¼ c fresh mushrooms, sliced		
¼ c sour cream		
¼ c cream cheese, softened		
½ c mozzarella cheese, shredded		
1 sprig fresh rosemary, chopped		

13. What is the scaling factor needed to scale tropical fruit kebabs from a total yield of 4 servings to a total yield of 100 servings?

14. Use the scaling factor found in the previous question to scale tropical fruit kebabs from a total yield of 4 servings to a total yield of 100 servings.

Tropical Fruit Kebabs (Yield: 100 servings)		
Original Yield ×	**SF** =	**Desired Yield**
2 c fresh pineapple, cut in 1 inch cubes		
1 c red seedless grapes		
2 c pears, cut in 1 inch pieces		
2 c mangoes, cut in 1 inch cubes		
4 kiwifruits, peeled and quartered		
2 tbsp lemon juice		
4 wooden skewers		

15. What is the scaling factor needed to scale California rolls from a total yield of 4 servings to a total yield of 100 servings?

16. Use the scaling factor found in the previous question to scale California rolls from a total yield of 4 servings to a total yield of 100 servings.

California Rolls (Yield: 100 servings)		
Original Yield ×	**SF** =	**Desired Yield**
¼ c rice vinegar		
1 tbsp sugar		
1 tsp salt		
1 c cooked sushi rice		
2 sheets nori, 8 x 7 inch sheets		
1 tbsp avocado, julienne cut		
1 tbsp cucumber, peeled, julienne cut		
½ c Alaskan king crab meat		
1 tbsp wasabi paste		
1 tbsp pickled ginger		

17. Create a descriptive hors d'oeuvres menu for the event.

Activity: Researching Garde Manger

Garde manger describes the kitchen station and the chef responsible for preparing cold pantry items such as salads, cheeses, and charcuterie. The garde manger may also be responsible for preparing garnishes, hors d'oeuvres, and appetizers.

Choose one of the following research assignments.

- Choose a country or region of a country and investigate the salads prepared in that area. Explain why certain ingredients and methods of preparation are used.
- Research olive oil and report on the different varieties. Explain the differences in oils resulting from the process used for production, or choose a particular country or region and explain the olive oil production method commonly used in that area. Explain why the oil has a particular taste difference in the region.
- Research one cheese from a region or country. Detail the history of the cheese as well as the common uses and changes that have occurred over time. If possible, bring in a sample of the cheese for the class to taste.
- Research a concept of organic farming for a particular farm, area, cheese, or vegetable. Include research detailing the impact of organic farming on health, economics, and marketing.
- Research the presentation styles used to showcase charcuterie items. Explain how to effectively use color, size, and shape of food items to enhance presentations.

1. Prepare a report and present the findings to the class. Use visuals and demonstrations to enhance the report. Include a list of sources used for the project.

Chapter 16 Review
Poultry, Ratites, and Related Game

Section 16.1 Poultry

Name ______________________________ **Date** ____________________

True-False

T F **1.** A capon is a surgically castrated male chicken that is less than eight months old and weighs approximately 4–7 lb.

T F **2.** Turkey is the most common type of poultry consumed.

T F **3.** Unlike other forms of poultry, ducks do not have white breast flesh.

T F **4.** Only mature, older pigeons are used in foodservice operations.

T F **5.** Guinea fowls are substantially leaner than chickens.

Multiple Choice

__________ **1.** Foie gras is the fattened ___ of a duck or goose.
A. gizzard
B. heart
C. kidney
D. liver

__________ **2.** A ___ turkey is a turkey that is less than eight months old and is 8–22 lb.
A. fryer
B. roaster
C. young
D. yearling

__________ **3.** A ___, also known as an African pheasant, is a farm-raised game bird that has lean, light red flesh.
A. quail
B. partridge
C. pigeon
D. young guinea fowl

__________ **4.** A Cornish hen is a ___ that is five to six weeks old and weighs 1½ lb or less.
A. turkey
B. chicken
C. pigeon
D. pheasant

_______________ **5.** ___ is a French term for meat that has been cooked and preserved in its own fat.
A. Confit
B. Cassoulet
C. Foie gras
D. Giblet

_______________ **6.** A ___ is a female chicken that has laid eggs for one or more seasons, is usually more than 10 months old, and weighs from 3–8 lb.
A. roaster
B. poussin
C. stewer
D. yearling chicken

_______________ **7.** Duck breasts are often seared in a sauté pan, finished in an oven, and served ___.
A. rare
B. medium-rare
C. medium
D. medium-well

_______________ **8.** A poussin is a very young chicken that weights ___ lb or less.
A. 1
B. 1½
C. 2
D. 2½

Completion

_______________ **1.** Only ___ geese are used in foodservice operations.

_______________ **2.** A(n) ___ is a young pigeon that is less than four weeks old and weighs approximately 1 lb or less.

_______________ **3.** A(n) ___ duckling is a duck that is less than 16 weeks old and weighs between 4–6 lb.

_______________ **4.** ___ is the collective term for various kinds of birds that are raised for human consumption.

_______________ **5.** A(n) ___ is a dish that consists of white beans stewed with duck fat, fresh sausage, and duck confit.

Chapter 16 Review
Poultry, Ratites, and Related Game

Section 16.2 Purchasing, Receiving, and Storing Poultry

Name ______________________ **Date** ______________

True-False

T F **1.** Purchasing whole poultry allows a chef to be more creative with the menu.

T F **2.** Leg and thigh flesh can be either dark or light.

T F **3.** For the most part, the only poultry sold to consumers is Grade A.

T F **4.** Older birds are desired over younger birds because of their tenderness and mild flavor.

T F **5.** The USDA inspection stamp indicates that the poultry was processed at a USDA-inspected plant.

Multiple Choice

______________ **1.** A ___ is the innermost section of a wing located between the first wing joint and the shoulder.
A. drumstick
B. drummette
C. wing paddle
D. wing tip

______________ **2.** The more a muscle is used, the ___ the muscle becomes.
A. less tender
B. less flavorful
C. lighter
D. darker

______________ **3.** A ___ consists of a tip, paddle, and drummette.
A. wing
B. tender
C. thigh
D. breast

______________ **4.** A ___ is the upper section of the leg located below the hip and above the knee joint.
A. half
B. wing
C. thigh
D. leg quarter

______________ **5.** Frozen poultry should be moved to a refrigeration unit ___ prior to use in order to thaw safely.
A. 2 hours
B. a day
C. five days
D. one week

Completion

______________ **1.** Giblets is the name for the grouping of the neck, heart, ___, and liver of a bird.

______________ **2.** A(n) ___ is the inner pectoral muscle that runs alongside the breastbone.

______________ **3.** Poultry should always be stored ___ other foods to prevent the juices from contaminating other foods.

______________ **4.** Frozen poultry should be stored in its original packaging at ___°F or below.

______________ **5.** To be classified as Grade ___, poultry must be free of deformities and have plump, meaty flesh and a thin layer of fat under the skin.

Matching

______________ **1.** Giblets

______________ **2.** Tender

______________ **3.** Drummette

______________ **4.** Wing tip

______________ **5.** Wing paddle

______________ **6.** Breast

______________ **7.** Tenderloin

______________ **8.** Drumstick

______________ **9.** Leg quarter

______________ **10.** Breast quarter

______________ **11.** Half

______________ **12.** Thigh

A. Lower portion of the leg located below the knee joint and above the ankle joint

B. Neck, heart, gizzard, and liver of a bird

C. Second section of a wing located between the two wing joints

D. A full half-length of a bird split down the breast and spine

E. Top front portion of the flesh above the rib cage

F. Half of a breast, a wing, and a portion of the back

G. Outermost section of a wing

H. Upper section of the leg located below the hip and above the knee joint

I. Small strip of breast

J. A thigh, a drumstick, and a portion of the back

K. Inner pectoral muscle that runs alongside the breastbone

L. Innermost section of a wing located between the first wing joint and the shoulder

Chapter 16 Review
Poultry, Ratites, and Related Game

Section 16.3 Fabrication of Poultry

Name ______________________________ Date ____________________

True-False

T F **1.** Boneless legs and thighs are never stuffed and roasted.

T F **2.** Poultry is commonly cut into halves, quarters, and eighths.

T F **3.** When cutting poultry into halves, the bird is split along both sides of the backbone from the neck to the tail.

T F **4.** Caul fat does not melt away during the cooking process and must be removed prior to service.

Multiple Choice

______________ **1.** Poultry that is cut into ___ yields two leg and thigh sections and two wing and breast sections.
- A. sixteenths
- B. eighths
- C. quarters
- D. halves

______________ **2.** ___ is a meshlike fatty membrane that can be wrapped around items to be roasted to add additional moisture and maintain a consistent shape.
- A. Connective tissue
- B. Caul fat
- C. Collagen
- D. Silverskin

______________ **3.** ___ is the most economical bird to fabricate.
- A. Chicken
- B. Turkey
- C. Duck
- D. Goose

______________ **4.** A(n) ___ is a skin-on, semiboneless breast that has the first bone of the wing still attached.
- A. roulade
- B. drummette
- C. wing
- D. airline poultry breast

Completion

______________________ **1.** Boneless ___ are the most popular poultry cut because of their versatility.

______________________ **2.** Each side of halved poultry yields a leg (drumstick and thigh), ___, and breast.

______________________ **3.** An airline breast makes an elegant presentation due to the exposed ___ bone.

______________________ **4.** ___ is the process of tying the legs and wings of a bird tightly to the body to keep a compact shape.

Name ______________________________ **Date** ________________

True-False

T F **1.** Barding is the process of laying a piece of fatback across the surface of a lean cut of meat to add moisture and flavor.

T F **2.** Marinades are always reserved and used as a sauce.

T F **3.** In the professional kitchen, it is common to stuff small birds such as Cornish hens and squabs, but not larger birds.

Multiple Choice

________________ **1.** Poultry is typically marinated ___ prior to cooking.
A. 5–10 minutes
B. 30–60 minutes
C. 2–3 days
D. 3–4 days

________________ **2.** When barding a bird, the fatback is often removed for ___ of the roasting process to allow browning.
A. a majority
B. the last 10 minutes
C. the last 30 minutes
D. the resting period

________________ **3.** Stuffing that is prepared and served separately is called ___.
A. trimmings
B. side plating
C. accompaniment
D. dressing

Completion

________________ **1.** A(n) ___ is a flavorful liquid used to soak uncooked foods such as meat, poultry, and fish to impart flavor and sometimes to tenderize.

_______________ **2.** Leftover marinade that came in contact with raw poultry is contaminated and must be ___.

_______________ **3.** As poultry cooks, the ___ is rendered from the fatback and absorbed into the flesh, which prevents the flesh from drying out and browning.

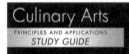

Chapter 16 Review
Poultry, Ratites, and Related Game

Section 16.5 Cooking Poultry

Name ______________________________ **Date** ____________________

True-False

T F **1.** The best cuts for sautéing are boneless breasts and medallions.

T F **2.** Legs and thighs have more connective tissue than breasts or wings, lending them to moist-heat cooking methods.

T F **3.** Duck breast is commonly served medium-rare.

T F **4.** When cooked properly, braised and stewed poultry should be tender and fall off the bone.

T F **5.** Very-low-temperature roasting is used for thick-skinned or fatty-skinned birds, such as ducks and geese.

T F **6.** Frying poultry produces a moist product because portioned cuts are typically breaded or battered to seal in the moisture before they are fried.

T F **7.** Birds larger than Cornish hens should be cut into quarters or eighths before grilling or broiling.

T F **8.** Braising is a combination cooking method that is often referred to as butter roasting.

Multiple Choice

______________ **1.** ___ is one of the most common methods for cooking whole poultry.
A. Sautéing
B. Stir-frying
C. Roasting
D. Steaming

______________ **2.** Testing for doneness using the touch, joints, or juices method is not recommended for ___ poultry.
A. grilled
B. barbequed
C. fried
D. sautéed

______________ **3.** ___ poultry while it is roasting adds moisture to the bird as it cooks.
A. Marinating
B. Seasoning
C. Salting
D. Basting

______________ **4.** Duck and turkey are the only types of poultry that are commonly ___.
A. poached
B. poêléed
C. smoked
D. sautéed

______________ **5.** Poultry continues to cook after it is removed from the heat due to ___.
A. carryover cooking
B. retrocooking
C. the Maillard reaction
D. convection

______________ **6.** ___ is a moist-heat cooking method in which food is cooked in a liquid that is held between 160°F and 180°F.
A. Poêléing
B. Poaching
C. Roasting
D. Sautéing

______________ **7.** A ___ is a thin piece of meat or poultry that is stuffed (filled), rolled, and cooked.
A. roulade
B. poêlé
C. drummette
D. kebab

______________ **8.** Chicken Marsala, chicken saltimbocca, and turkey piccata are ___ dishes.
A. stewed
B. fried
C. braised
D. sautéed

Completion

______________ **1.** The four methods that chefs use to determine the doneness of poultry are temperature, touch, ___, and juices.

______________ **2.** ___ and stewing are used to prepare tough or less flavorful birds that benefit from cooking in a flavorful liquid for a long period.

______________ **3.** Juices from raw poultry are red, and juices from fully cooked birds are ___.

______________ **4.** When checking poultry for doneness, the thermometer should be inserted close to, but not touching, any large ___.

______________ **5.** Regardless of the cooking method used, poultry should be cooked to an internal temperature of ___°F, except in the case of duck.

______________ **6.** Prior to being sautéed, poultry is commonly ___ in seasoned flour to help seal in moisture, promote even browning, and allow the seasonings to stick to the flesh.

Chapter 16 Review
Poultry, Ratites, and Related Game

Section 16.6 Ratites and Game Birds

Name ______________________________ **Date** ________________

True-False

T F **1.** The flavor of ratite flesh is similar to beef, but sweeter.

T F **2.** Farm-raised game birds are processed at USDA-inspected facilities and are graded by the USDA.

T F **3.** Quail eggs are not served in restaurants due to their small size.

T F **4.** Female pheasants are preferred over male pheasants because female pheasants are more tender and moist.

T F **5.** The flesh of farm-raised wild turkeys is all white.

Multiple Choice

______________ **1.** A ratite is any of a large variety of flightless birds that have flat breastbones and small ___ in relation to their body size.

A. thighs
B. legs
C. wings
D. breasts

______________ **2.** A ___ is the smallest game bird.

A. quail
B. pheasant
C. grouse
D. partridge

______________ **3.** A(n) ___ is the largest ratite and can weigh 300–400 lb.

A. emu
B. peacock
C. rhea
D. ostrich

______________ **4.** Rheas are native to ___.
A. South America
B. Asia
C. Africa
D. Europe

______________ **5.** A ___ is a game bird that resembles a chicken in appearance but has thicker, stronger legs and dark flesh.
A. quail
B. grouse
C. wild turkey
D. partridge

Completion

______________ **1.** The fan, taken from a muscle in the ___, is the most tender cut of ratite.

______________ **2.** A(n) ___ is the second largest ratite and can weigh between 125 – 140 lb.

______________ **3.** ___ turkey flesh is tougher, leaner, less moist, and less meaty than domesticated turkey flesh.

______________ **4.** A(n) ___ bird is a wild bird that is hunted for human consumption.

______________ **5.** A(n) ___ is a game bird that weighs about 1 lb, has white flesh, and yields an edible portion for two people.

Chapter 16 Activities
Poultry, Ratites, and Related Game

Name ______________________________ **Date** ____________________

Activity: Identifying Market Forms of Poultry

Poultry is available in a variety of market forms, including whole birds and fabricated cuts. Knowledge of these market forms is necessary for accurate ordering.

Match each of the following market forms of poultry to its image.

_______________ **1.** Breast

_______________ **2.** Ground

_______________ **3.** Thighs

_______________ **4.** Liver

_______________ **5.** Wings

_______________ **6.** Neck

_______________ **7.** Legs

_______________ **8.** Whole

_______________ **9.** Heart and gizzard

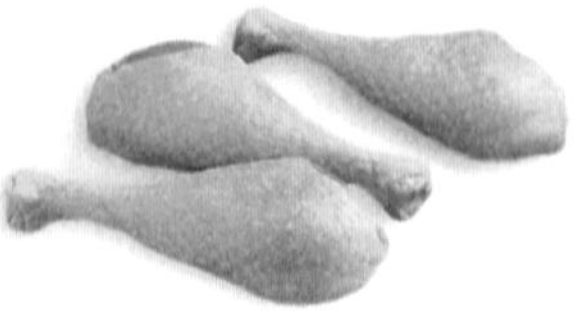
National Turkey Federation
Ⓐ

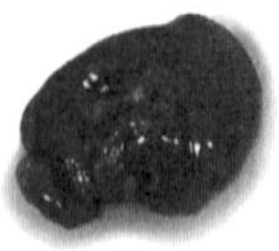
Ⓑ

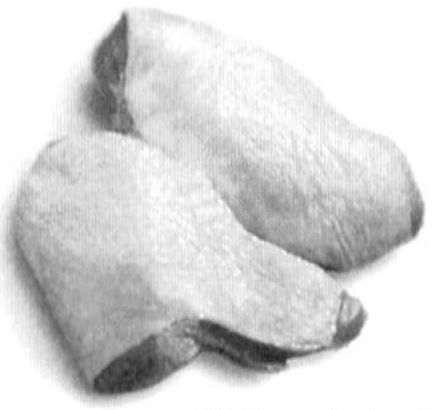
National Turkey Federation
Ⓒ

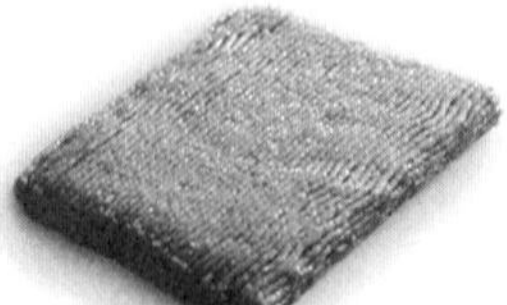
National Turkey Federation
Ⓓ

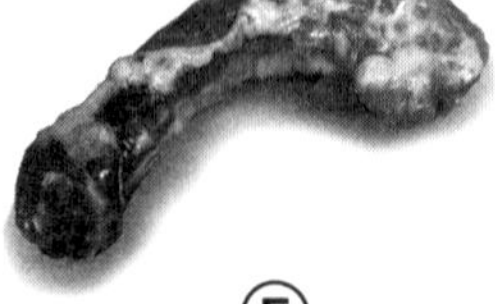
Ⓔ

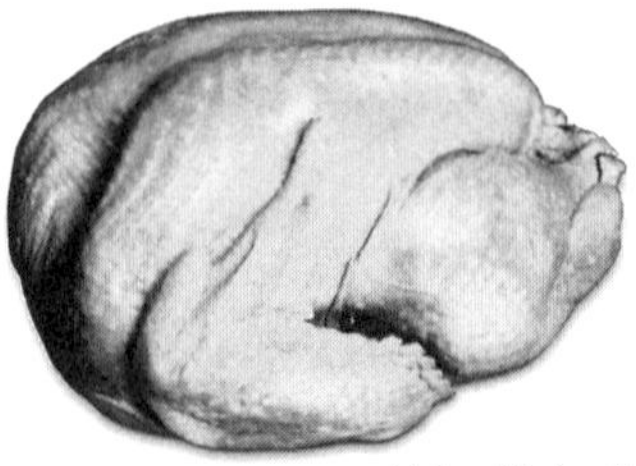
National Turkey Federation
Ⓕ

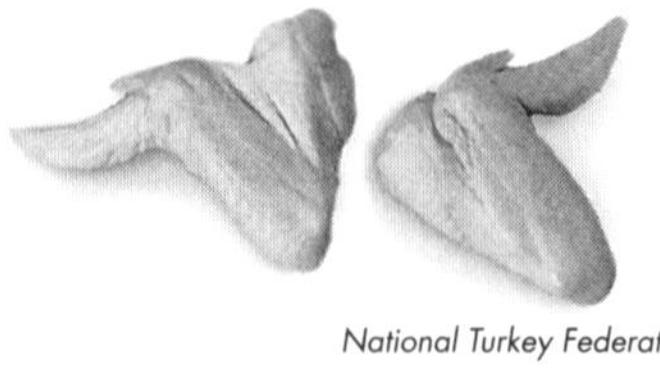
National Turkey Federation
Ⓖ

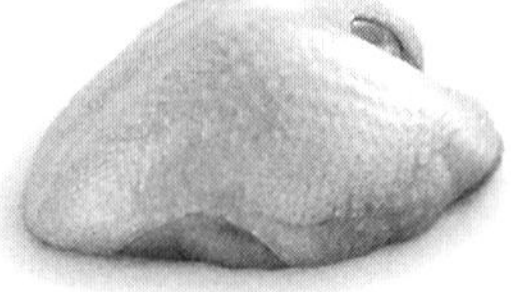
National Turkey Federation
Ⓗ

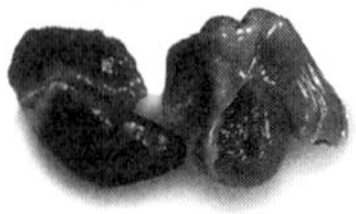
Ⓘ

Activity: Identifying Poultry Cut Points

Fabricating techniques can be applied to almost any type of poultry because all types have similar bodies and bone structures. Understanding the skeletal structure of a chicken can aid the fabrication process.

Answer the following items by identifying the letter that corresponds to the cut point.

__________ **1.** Front half of back from back half of back

__________ **2.** Drumstick from foot

__________ **3.** Neck from breast/back

__________ **4.** Thigh from drumstick

__________ **5.** Leg from back

__________ **6.** Wing from breast

__________ **7.** Breast from back

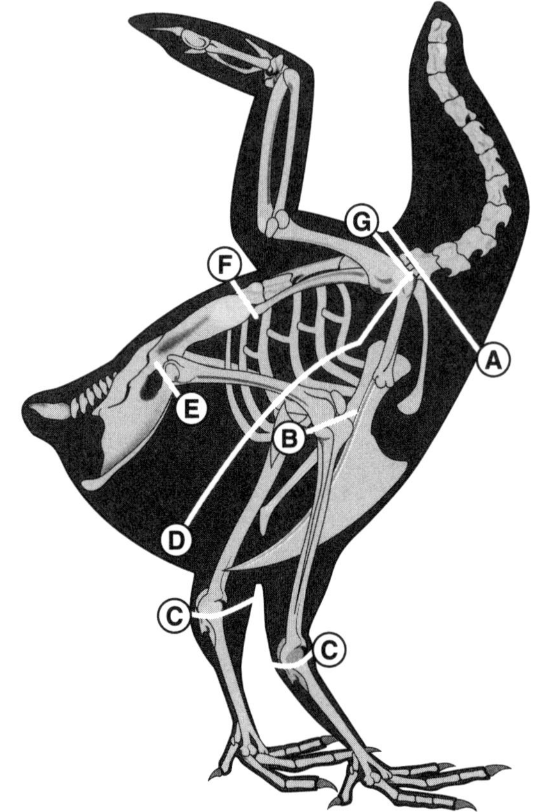

Activity: Fabricating Poultry

There are many ways that poultry can be cut into portions. Typical fabrication methods for poultry include cutting it into halves, quarters, and eighths. Poultry may also be fabricated to produce boneless breasts, airline breasts, a boned leg and thigh, and a boned whole chicken.

Perform the following activities using chicken. Have an instructor check the result and sign below when complete.

1. Truss a chicken.

Trussing Proficiency

Instructor:____ Date:_______________

2. Cut a chicken in half.

Fabricating into Halves Proficiency

Instructor:____ Date:_______________

3. Cut a chicken into quarters.

Fabricating into Quarters Proficiency

Instructor:____ Date:_______________

4. Cut a chicken into eighths.

Fabricating into Eighths Proficiency

Instructor:____ Date:_______________

5. Cut boneless breasts.

Fabricating Boneless Breasts Proficiency

Instructor:____ Date:_______________

6. Cut airline breasts.

Fabricating Airline Breasts Proficiency

Instructor:______________________________________ Date:______________

7. Bone out a leg and thigh.

Boning a Leg and Thigh Proficiency

Instructor:______________________________________ Date:______________

8. Bone a whole chicken.

Boning a Whole Chicken Proficiency

Instructor:______________________________________ Date:______________

Activity: Preparing Poultry

Chicken should be cooked to an internal temperature of 165°F for a minimum of 15 seconds.

Roast two fryers, trussing one bird and leaving the other fryer untrussed.

1. Describe the appearance, texture, and flavor of the trussed fryer.

2. Describe the appearance, texture, and flavor of the fryer that was not trussed.

Broil or grill a boneless, skinless chicken breast, a bone-in chicken breast with skin, and a bone-in chicken breast with skin that has been marinated.

3. Describe the appearance, texture, and flavor of the boneless, skinless chicken breast.

4. Describe the appearance, texture, and flavor of the bone-in chicken breast with skin.

5. Describe the appearance, texture, and flavor of the bone-in chicken breast with skin that has been marinated.

Stew or braise a fryer, a roaster, or a stewer.

6. Describe the appearance, texture, and flavor of the chicken.

Chapter 17 Review
Fish, Shellfish, and Related Game

Section 17.1 Fish

Name ______________________________ **Date** ______________

True-False

T F **1.** Sharks, skates, and stingrays are cartilaginous fish.

T F **2.** Fatty fish, such as salmon, are best prepared using moist-heat cooking methods.

T F **3.** The fat content of a fish can affect both flavor and the cooking method required.

T F **4.** Flatfish are the most common type of fish.

T F **5.** Fish are considered lean or fatty based on the thickness of their connective tissue.

Multiple Choice

______________ **1.** Fish are most often grouped by ___ and by the composition of their external shape and structure.
- A. color
- B. approximate weight
- C. reproductive habits
- D. habitat

______________ **2.** A ___ is any thin, wide fish with both eyes located on one side of the head and a backbone that runs from head to tail through the midline of the body.
- A. roundfish
- B. flatfish
- C. cartilaginous
- D. bony

______________ **3.** ___ often have a smooth, tough outer skin without scales.
- A. Cartilaginous fish
- B. Roundfish
- C. Fatty fish
- D. Flatfish

______________ **4.** ___ is an example of a lean fish.
- A. Catfish
- B. Rainbow trout
- C. Pacific cod
- D. Mackerel

______________ **5.** ___ is an example of a flatfish.
A. Flounder
B. Trout
C. Skate
D. Salmon

Completion

______________ **1.** A(n) ___ is any fish with a cylindrical body, an eye located on each side of the head, and a backbone that runs from head to tail in the center of the body.

______________ **2.** ___ swim parallel to the surface of the water with one side facing down and the side having both eyes facing toward the surface.

______________ **3.** A(n) ___ fish is any fish that has a skeleton composed of cartilage instead of bones.

______________ **4.** The ___ side of a flatfish is light in color.

______________ **5.** A(n) ___ is any of a classification of animal that has fins for moving through the water, gills for breathing, and an internal skeleton made of bones or cartilage.

Chapter 17 Review
Fish, Shellfish, and Related Game

Section 17.2 Types of Fish

Name ______________________________ **Date** ______________________

True-False

T F **1.** Common examples of anadromous fish include trout, perch, catfish, tilapia, and smelt.

T F **2.** Sole is a lean, saltwater flatfish with pale-brown skin on the top side.

T F **3.** Haddock cod, also known as grey cod, is most often marketed as "true cod."

T F **4.** Only the tail section and liver of a monkfish are edible.

T F **5.** Each wing of a skate produces two separate fillets.

Multiple Choice

______________ **1.** A ___ is a fatty roundfish with a slender body, pointed head, and deeply forked tail.
A. smelt
B. tilapia
C. walleye
D. trout

______________ **2.** A ___ is a lean saltwater roundfish that has colorful skin and firm, pink flesh.
A. barramundi
B. herring
C. mackerel
D. mahi-mahi

______________ **3.** ___ sturgeon are the most prized variety of sturgeon because of their caviar and firm, oily flesh.
A. Osetra
B. Sevruga
C. Beluga
D. Lake

______________ **4.** A sardine is a very small young ___.
A. cod
B. herring
C. halibut
D. sturgeon

______________ **5.** A ___ is a lean, freshwater roundfish native to the Great Lakes and northern Canada.

A. perch
B. trout
C. catfish
D. flounder

Completion

______________ **1.** ___ is a process of raising fish, shellfish, or related game in a controlled inland environment.

______________ **2.** A(n) ___ fish is a fish that begins life in freshwater, spends most of its life in saltwater, and returns to freshwater to spawn.

______________ **3.** Bigeye and yellowfin tuna are both marketed as ___ tuna, which is most often seared and served rare or medium-rare.

______________ **4.** A distinguishing factor of red snapper is that its fillets do not ___ when cooked.

______________ **5.** A(n) ___ is a fatty cartilaginous fish with a long, slender body that resembles a snake.

Chapter 17 Review
Fish, Shellfish, and Related Game

Section 17.3 Purchasing, Receiving, and Storing Fish

Name ______________________________ **Date** ______________

True-False

T F **1.** Frozen fish cannot be purchased whole.

T F **2.** Type 3 inspection of fish involves continual inspection of the fresh product from the time it arrives at the processing plant to the moment it is packaged for sale.

T F **3.** Only fish that is processed under a voluntary Type 1 inspection is eligible for grading.

T F **4.** Fish that has been refrozen will have a poor texture.

T F **5.** A pan-dressed fish is a cross section of a dressed fish.

Multiple Choice

______________ **1.** A ___ fish is a fish that has had only the viscera removed.
A. whole
B. drawn
C. dressed
D. pan-dressed

______________ **2.** Fish with ___ or sunken eyes should be rejected.
A. round
B. slightly bulging
C. cloudy
D. clear

______________ **3.** ___ is a brine-cured, cold-smoked salmon that is less salty than lox.
A. Sardine
B. Gravlax
C. Nova
D. Herring

______________ **4.** Fresh fish can be stored a maximum of ___.
A. one to two days
B. five to six days
C. one to two weeks
D. one month

______________ **5.** ___ contain ungraded fish fillets that are frozen in packets, typically one to three fillets per packet, wrapped in cellophane, frozen, and packaged six packets per box.

A. Individually quick-frozen boxes
B. Block-frozen fish
C. Layer packs
D. Cello packs

Completion

______________ **1.** Optional inspections of fresh fish are carried out by the ___, which is part of the United States Department of Commerce.

______________ **2.** Fresh fish should have an internal temperature of ___°F or lower to be accepted at delivery.

______________ **3.** ___ is a designation for products preserved using a method in which each item is glazed with a thin layer of water and frozen individually.

______________ **4.** A fish ___ is a cross section of a dressed fish.

______________ **5.** ___ is the process of covering an item with water to form a protective coating of ice before the item is frozen.

Chapter 17 Review
Fish, Shellfish, and Related Game

Section 17.4 Fabrication of Fish

Name ______________________________ **Date** ________________

True-False

T F **1.** To scale a fish, the head of the fish is held firmly with the guiding hand and a fish scaler is used to scrape against the scales down toward the tail.

T F **2.** To cut a roundfish into steaks, the fish is sliced vertically into equally sized pieces.

T F **3.** Scales must be removed from fish prior to preparation.

Multiple Choice

______________ **1.** To skin a fillet, start at the ___.
- A. tail
- B. fin
- C. head
- D. belly

______________ **2.** To fillet a flatfish, a ___ knife is used to cut along the backbone from the head to the tail.
- A. chef's
- B. paring
- C. bread
- D. boning

______________ **3.** ___ loins commonly have a higher fat content than other loins and are therefore more desirable for sashimi.
- A. Bottom side
- B. Topside
- C. Belly
- D. Back

______________ **4.** ___ are used to remove pinbones from a roundfish fillet.
- A. Boning knives
- B. Needle-nose pliers
- C. Chef's knives
- D. Tongs

Completion

______________ **1.** A(n) ___ is the most common serving portion from a fish.

______________ **2.** Flatfish yield ___ fillets.

______________ **3.** The ___ of a flatfish runs through the midline of the fish.

Name ______________________________ Date ________________

True-False

T F **1.** When the thickest part of a thin fish is prodded with a fork, flesh that is done will be opaque and the juices will be clear.

T F **2.** Steaming adds no additional calories or fat to a fish.

T F **3.** Fish are typically kept whole for frying.

T F **4.** A fish that is done will have firm and dense interior edges and a moist, opaque center.

Multiple Choice

________________ **1.** ___ uses a small amount of hot fat to sear and cook the fish.
A. Poaching
B. Frying
C. Grilling
D. Sautéing

________________ **2.** To ensure that the fish cooks properly and does not become tough, it is essential that a poaching liquid remains below a simmer and between ___.
A. 60°F and 80°F
B. 100°F and 120°F
C. 160°F and 180°F
D. 200°F and 220°F

________________ **3.** An instant-read thermometer that registers ___°F when inserted into the thickest part of a thick-fleshed fish indicates the fish is cooked through.
A. 95
B. 120
C. 145
D. 185

Completion

________________ **1.** En papillote is a technique in which food is steamed in a(n) ___ as it bakes in an oven.

____________________ **2.** When cooking fish fillets with the skin on, the skin should be ___ with a sharp knife first so that the fillet does not curl during cooking.

____________________ **3.** A lean fish may be broiled or baked if it is ___ during the cooking process.

Name ______________________________ Date ________________

True-False

T F **1.** Shellfish are commonly categorized as crustaceans or mollusks.

T F **2.** All mollusks have a hard, external shell.

T F **3.** An oyster is a type of mollusk that has a hard, external shell.

Multiple Choice

______________ **1.** A(n) ___ is a shellfish with a soft, nonsegmented body.
A. crustacean
B. mollusk
C. shrimp
D. cephalopod

______________ **2.** A(n) ___ is an example of a mollusk that does not have an external shell.
A. squid
B. oyster
C. mussel
D. clam

______________ **3.** The hard external shell of a shellfish functions as a ___.
A. muscle
B. digestive system
C. nervous system
D. skeleton

Completion

______________ **1.** ___ is the classification of aquatic invertebrates that may or may not have a hard, external shell.

______________ **2.** A(n) ___ is a shellfish that has a hard, segmented shell that protects its soft flesh and does not have an internal bone structure.

______________ **3.** The hard, external shell of a shellfish is called a(n) ___.

Section 17.7 Types of Shellfish

Name ______________________________ **Date** ________________

True-False

T F **1.** Crustaceans include shrimp, prawns, lobsters, crayfish, and crabs.

T F **2.** The soft green substance found in the body cavity of a lobster is the liver and pancreas, which are collectively called the roe.

T F **3.** Most inedible crabs have five pairs of legs.

T F **4.** The whisker-like threads that extend outside the shell of a mussel, referred to as a "beard," allow the mussel to attach to items for protection.

T F **5.** Octopus flesh is white, firm, and sweet.

T F **6.** Squids are the only type of cephalopod that can change their skin color at will.

Multiple Choice

________________ **1.** A ___ lobster is a large lobster with a dark bluish-green shell, two large, heavy claws, and four slender legs on each side of its body.
A. rock
B. spiny
C. Maine
D. Norway

________________ **2.** Blue crab is a(n) ___ crab with blue claws and a dark blue-green, oval shell.
A. North American
B. South American
C. Pacific
D. Atlantic

________________ **3.** A ___ is a lobster that is dying.
A. crayfish
B. tomalley
C. prawn
D. sleeper

_______________ **4.** A ___ crab is an Atlantic crab with a brownish-red shell and large claws of unequal size.
A. king
B. stone
C. blue
D. Dungeness

_______________ **5.** ___ mussels are the most common variety of edible mussel.
A. Blue
B. Atlantic
C. Surf
D. Greenlip

_______________ **6.** ___ crabs are one of the few edible crabs that have four pairs of legs.
A. Blue
B. King
C. Snow
D. Dungeness

Completion

_______________ **1.** In the culinary industry, the terms shrimp and ___ are used interchangeably.

_______________ **2.** A(n) ___ is a translucent cephalopod with two tentacles, eight sucker-equipped arms, an internal shell called a cuttlebone, and large eyes at the base of its head.

_______________ **3.** The ___ muscle is a muscle that opens and closes the shell of a bivalve.

_______________ **4.** ___ varieties include abalone and conch.

_______________ **5.** ___ scallops and sea scallops are the two primary varieties of scallops.

_______________ **6.** A(n) ___ is a freshwater crustacean that resembles a tiny lobster and has a similar flavor to that of shrimp.

Matching

_______________ **1.** Siphon

_______________ **2.** Adductor muscle

_______________ **3.** Univalve

_______________ **4.** Cephalopod

_______________ **5.** Bivalve

A. Tubular organ that is used to draw in or eject fluids

B. Any of a variety of mollusks that do not have an external shell

C. Mollusk with a top shell and a bottom shell connected by a central hinge that can close for protection

D. Muscle that opens and closes the shell of a bivalve

E. Mollusk that has a single solid shell and a single foot

SECTION 17.8 PURCHASING, RECEIVING, AND STORING SHELLFISH

Name ______________________________ **Date** ____________________

True-False

T F **1.** Surimi is a fish product made from a mixture of fish and/or shellfish and other ingredients.

T F **2.** Shrimp packed and labeled 21/25 count indicates that there is an average of 21 – 25 shrimp per 1 lb package.

T F **3.** Clams, oysters, and scallops are never sold with the shell removed.

T F **4.** After receiving shellfish that is thawed on the edges, it should be refrozen.

T F **5.** When live shellfish are packed by count, a higher count indicates a smaller size.

Multiple Choice

__________ **1.** Live crustaceans should be covered with wet ___ to prevent them from drying out.

A. plastic wrap
B. sand
C. seaweed
D. styrofoam

__________ **2.** Frozen shellfish is commonly packaged by ___.

A. size
B. count
C. weight
D. shape

__________ **3.** Once accepted, it is imperative to store shellfish at ___°F or below and to use it as quickly as possible to maintain quality.

A. 0
B. 34
C. 45
D. 90

__________ **4.** In the absence of a tank, live crustaceans live ___.

A. four to six hours
B. one to two days
C. two to four days
D. one to two weeks

_______________ **5.** If handled in the proper manner, oysters remain fresh for up to ___.
A. one week
B. two weeks
C. one month
D. two months

Completion

_______________ **1.** All mollusks must be delivered with a(n) ___ listing the dealer's contact information and identification number, the original harvester, the harvest date and general area, the type and quantity of shellfish, a 90 day retention notice, and a consumer advisory.

_______________ **2.** If a shell is open and does not ___ when handled, the oyster or clam is dead and should be discarded.

_______________ **3.** Oyster and clam shells that are unusually heavy probably contain ___ and should also be discarded.

_______________ **4.** Surimi looks, cooks, and tastes like ___.

_______________ **5.** Maximum shelf life is obtained by storing frozen shellfish at ___°F or below.

Chapter 17 Review
Fish, Shellfish, and Related Game

Section 17.9 Fabrication of Shellfish

Name ______________________________ **Date** ____________________

True-False

T F **1.** Mussels have hairlike threads called beards that need to be removed prior to preparation.

T F **2.** If a squid is dark in color and the eyes are intact, it has already been cleaned.

T F **3.** Shucking is the process of opening a univalve.

T F **4.** When splitting a lobster tail, use kitchen shears to cut down the center of the shell, starting at the head end.

T F **5.** To shuck clams, the blade of an oyster knife is inserted between the top shell and bottom shell until halfway inserted.

Multiple Choice

______________ **1.** Cleaning soft-shell crabs involves removing the inedible portions, such as the eyes and the ___.

A. legs
B. claws
C. beard
D. apron

______________ **2.** When cleaning octopuses, the ___ are not discarded.

A. beaks
B. arms
C. heads
D. skins

______________ **3.** Clams may be ___ open to access the flesh.

A. fried
B. boiled
C. steamed
D. broiled

______________ **4.** If a squid is ___, it must be cleaned before being prepared.

A. dark in color
B. missing its eyes
C. creamy white
D. fishy smelling

______________ **5.** When cleaning squid, hold the body with the guiding hand and pull out the ___ (transparent backbone) with the other hand.
A. cuttlebone
B. pen
C. exoskeleton
D. scapula

Completion

______________ **1.** ___ contain a sand vein, or intestinal tract, that must be removed prior to further preparation.

______________ **2.** Broiled lobster tail is a preparation that requires the shell of the tail to be ___.

______________ **3.** When shucking oysters, slide the oyster knife under the ___ muscle to separate the flesh from the shell.

______________ **4.** The ___ is located tucked under the body of soft shell crabs and is used to carry and conceal eggs.

______________ **5.** If the ___ of a mussel is removed too soon, the mussel will die.

Chapter 17 Review
Fish, Shellfish, and Related Game

Section 17.10 Cooking Shellfish

Name ______________________________ **Date** ____________________

True-False

T F **1.** Shrimp, prawns, and crayfish are often sautéed, fried, or steamed.

T F **2.** Scallops are typically sautéed to produce a golden exterior.

T F **3.** Shellfish require short cooking times and overcook quickly.

T F **4.** Octopus cannot be used to make sushi.

T F **5.** Oysters are always baked with a topping and never served raw on the half shell.

Multiple Choice

______________ **1.** A lobster shell turns ___ when it is done cooking.
A. red
B. black
C. orange
D. brown

______________ **2.** ___ is a good indicator of the doneness of shellfish.
A. Smell
B. Color
C. Shape
D. Texture

______________ **3.** The center of a scallop turns ___ when it is done.
A. reddish-brown
B. opaque
C. slightly pink
D. solid white

______________ **4.** If octopus is overcooked, it becomes ___.
A. buttery
B. creamy
C. rubbery
D. mushy

_______________ 5. An oyster po'boy sandwich includes ___ oysters.
A. sautéed
B. steamed
C. roasted
D. fried

Completion

_______________ 1. The liquid used for poaching whole lobsters should be ___°F.

_______________ 2. Shrimp cure and turn a slightly ___ color when cooked.

_______________ 3. Squid may be cut into rings and then quickly sautéed or lightly breaded and fried to make ___.

_______________ 4. When preparing squid, the ___ sac is often used as a black food coloring.

_______________ 5. Precooked crab legs only need to be ___ through prior to serving.

Name ______________________________ Date ________________

True-False

T F **1.** The two most common varieties of edible turtles are green turtles and snapping turtles.

T F **2.** Turtle flesh has a different flavor depending on which part of the turtle it comes from.

T F **3.** Snake flesh is similar in color, flavor, and texture to veal.

T F **4.** The hind legs of bullfrogs contain a large amount of dark flesh.

T F **5.** Alligator leg and jaw flesh is sold as smaller pieces that can be breaded, fried, and sold as "alligator bites."

Multiple Choice

__________ **1.** Escargot is the French term for ___.
A. frog
B. turtle
C. snail
D. alligator

__________ **2.** Turtle flesh is most commonly used to make turtle ___.
A. desserts and appetizers
B. fillets and rings
C. steaks, chops, and roasts
D. soup, stews, and gumbos

__________ **3.** Farm-raised alligators are fed a natural diet of ___.
A. fish and small rodents
B. fish and frogs
C. rodents and small animals
D. cephalopods and mussels

__________ **4.** Snapping turtle ___ are dark flesh.
A. legs
B. necks
C. back straps
D. bellies

______________________ **5.** Frog legs are sold by the pair, with the most desirable legs averaging ___ pairs per pound.

A. 2–3

B. 5–6

C. 8–10

D. 12–15

Completion

______________________ **1.** ___ is the most common variety of snake raised for human consumption.

______________________ **2.** The best frog legs come from ___ raised on frog farms.

______________________ **3.** To help ensure snails are safe for eating, their stomachs are ___.

______________________ **4.** ___ turtles have bumpy shells with sharp, pointed tips and sharp beaklike mouths.

______________________ **5.** Alligators are raised for two cuts of flesh: the tenderloin and the ___.

Chapter 17 Activities
Fish, Shellfish, and Related Game

Name ______________________________ **Date** ____________________

Activity: Identifying Market Forms of Fish

Fish may be purchased fresh, frozen, or processed. The more that is done to a fish prior to delivery, the higher the cost per pound. Fish are commonly sold whole, drawn, dressed or pan-dressed, as steaks, and as fillets.

Match each of the following market forms of fish to the correct image.

____________ **1.** Fillet

____________ **2.** Pan-dressed

____________ **3.** Drawn

____________ **4.** Steaks

____________ **5.** Butterfly fillet

____________ **6.** Whole

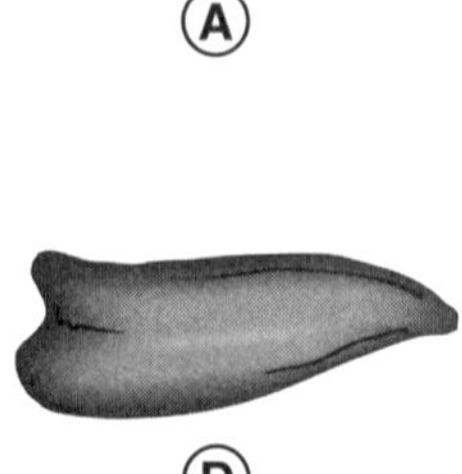

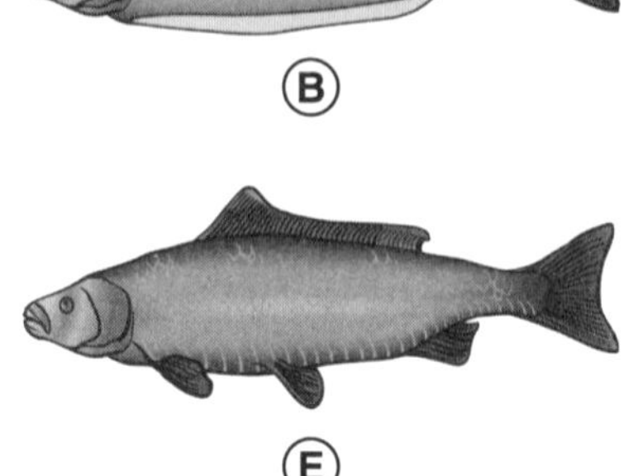

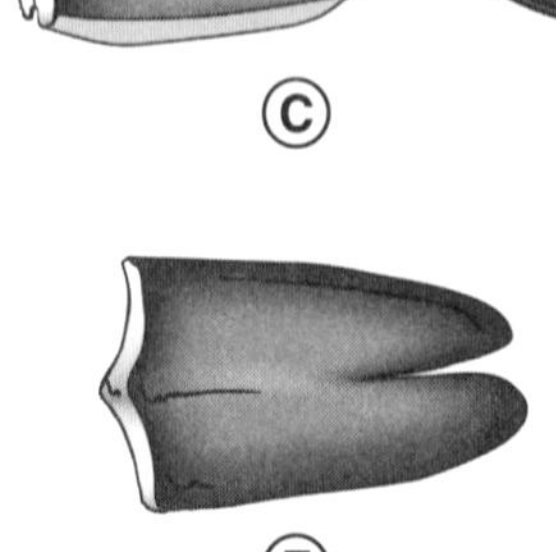

D

E

F

Activity: Determining Freshness of Fish

Fresh fish spoils rapidly, so it is essential to check the smell, appearance, and flesh of the fish upon receipt.

Complete the following.

1. What is an acceptable smell for a fresh fish?

2. What is an acceptable appearance for the eyes of a fresh fish?

3. What is an acceptable appearance for the gills of a fresh fish?

4. What is an acceptable appearance for the fillets of a fresh fish?

5. When a fresh fish is touched, how should the exterior surface feel?

6. When a fresh fish is touched, how should the scales feel?

7. When a fresh fish is touched, how should the fins feel?

8. When a fresh fish is touched, how should the flesh feel?

Activity: Preparing Fish and Cephalopods

Most fish, shellfish, and related game are naturally tender and require very little cooking time. Some fish tend to become dry when certain cooking methods are applied.

Broil a lean fish and a fatty fish. Respond to the following items.

1. Describe the appearance, texture, and flavor of the lean fish.

2. Describe the appearance, texture, and flavor of the fatty fish.

Sauté a lean fish and a fatty fish. Respond to the following items.

3. Describe the appearance, texture, and flavor of the lean fish.

4. Describe the appearance, texture, and flavor of the fatty fish.

5. What conclusions can be drawn about cooking methods for lean and fatty fish?

Prepare the fried calamari with lemon butter recipe found in Chapter 17: Fish, Shellfish, and Related Game. Respond to the following items.

6. Describe the appearance, texture, and flavor of the fried calamari.

7. What conclusions can be drawn about frying cephalopods?

Activity: Identifying Fish and Shellfish

A chef must be able to identify fish and shellfish by appearance. The appearance and smell of fish and shellfish are also important in determining freshness and quality of ingredients.

Match each fish and shellfish to its image.

______ **1.** Squid
______ **2.** Snow crab
______ **3.** Monkfish
______ **4.** Striped bass
______ **5.** Shrimp
______ **6.** Cuttlefish
______ **7.** Red snapper
______ **8.** Skate
______ **9.** Lake trout
______ **10.** Bluefin tuna
______ **11.** King salmon
______ **12.** Sole
______ **13.** Cod
______ **14.** Swordfish
______ **15.** Catfish

National Oceanic and Atmospheric Administration
Ⓐ

U.S Geological Survey
Ⓑ

Ⓒ

Plitt Seafood
Ⓓ

Ⓔ

Ⓕ

Ⓖ

Ⓗ

Ⓘ

National Oceanic and Atmospheric Adminsistration
Ⓙ

Ⓚ

Plitt Seafood
Ⓛ

Ⓜ

Ⓝ

U.S. Fish and Wildlife Service
Ⓞ

Activity: Fabricating Fish and Shellfish

Fish may be purchased whole and then fabricated in-house or may be purchased in a portion-controlled or processed form. Knowing and using proper fabrication techniques for fish are important in the professional kitchen.

Perform the following procedures. Have an instructor check the result and sign below when complete.

1. Scale a fish.

Scaling Proficiency

Instructor:____ Date:_________________

2. Cut a roundfish into steaks.

Cutting a Roundfish into Steaks Proficiency

Instructor:____ Date:_________________

3. Fillet a roundfish.

Filleting a Roundfish Proficiency

Instructor:____ Date:_________________

4. Skin a fillet.

Skinning a Fillet Proficiency

Instructor:____ Date:_________________

5. Fillet a flatfish.

Filleting a Flatfish Proficiency

Instructor:____ Date:_________________

6. Devein shrimp.

Deveining Shrimp Proficiency

Instructor:____ Date:_________________

7. Split a lobster tail.

Splitting a Lobster Tail Proficiency

Instructor:____ Date:________________

8. Shuck oysters.

Shucking Oysters Proficiency

Instructor:____ Date:________________

9. Shuck clams.

Shucking Clams Proficiency

Instructor:____ Date:________________

Activity: Determining Local Fish Availability

The availability of fish often depends on the location of the foodservice operation. Knowing what types of fish are available locally can inspire new and enticing menu offerings.

Visit a local grocery store or market to determine the availability of fish. Complete the following chart.

Local Fish Availability				
Name	Condition*	Market Form†	Preparation Required	AP Cost
1.				$______ per lb
2.				$______ per lb
3.				$______ per lb
4.				$______ per lb
5.				$______ per lb
6.				$______ per lb
7.				$______ per lb
8.				$______ per lb
9.				$______ per lb
10.				$______ per lb
11.				$______ per lb
12.				$______ per lb
13.				$______ per lb
14.				$______ per lb
15.				$______ per lb

* live, fresh frozen, canned, or smoked
† whole, drawn, dressed, steaks, fillets, butterfly fillets

Activity: Researching the Fishing Industry

The fishing industry encounters many issues that affect the day-to-day operations of fisheries and impact the fishing market. Certain issues are local and others affect the industry worldwide. Some of these issues include the following:

- Ecological problems and concerns
- Endangered fish and sustainability
- Fishing rules and regulations
- Parasites
- Fish toxins
- Fish waste recycling

Choose an issue that affects the fishing industry and use the following outline to write an essay about the issue.

I. Introduction
   A. What is the issue?
II. Background information
   A. When did this issue emerge?
   B. What caused it to become an issue?
   C. How has this issue affected the fishing industry?
   D. How has this issue affected consumers?
III. Current situation
   A. What is being done about the issue?
   B. Who is involved?
   C. What legal and/or ethical implications are involved with this issue?
IV. Possible solutions
   A. What are the short-term effects of the possible solutions on the fishing industry and consumers?
   B. What are the long-term effects of the possible solutions on the fishing industry and consumers?
V. Conclusion

1. Prepare a report and present the findings to the class using the outline provided. Use visuals such as pictures, charts, and graphs to enhance the report. Include a list of the sources used for the project.

Chapter 18 Review
Beef, Veal, and Bison

Section 18.1 Beef

Name ______________________________ Date ______________

True-False

T F 1. Grain-finishing increases the marbling of the meat from a steer or heifer.

T F 2. Types of domesticated cattle include steers, heifers, cows, and bulls.

T F 3. Elastin is left on the meat while cooking.

Multiple Choice

______________ 1. Grass-fed beef is beef from cattle that were raised on grass for ___.
A. two months before harvest
B. one years
C. two years
D. their entire lives

______________ 2. Steers and heifers produce the best-quality beef and are typically ___ old when they are harvested.
A. 4–6 weeks
B. 5–15 weeks
C. 8–24 months
D. over 2 years

______________ 3. Grain-fed beef is beef from cattle that were grass-fed and then grain-finished in a feedlot for approximately ___.
A. 10–15 days
B. 4–6 months
C. 8–12 months
D. 1–2 years

Completion

______________ 1. Beef is the flesh of domesticated ___.

______________ 2. A(n) ___ is a male calf that has been castrated prior to reaching sexual maturity.

______________ 3. ___ fat is the fat that runs between muscles in a cut of meat.

Matching

_______________ **1.** Marbling

_______________ **2.** Collagen

_______________ **3.** Heifer

_______________ **4.** Fat cap

_______________ **5.** Elastin

A. A female calf that has not had a calf of her own

B. Tough, rubbery, silver-white connective tissue that does not break down when heated

C. Soft, white, connective tissue that breaks down into gelatin when heated

D. Fat found within a muscle

E. Fat that surrounds the outside of a muscle on a cut of meat

Chapter 18 Review
Beef, Veal, and Bison

Section 18.2 Purchasing Beef

Name ______________________________ **Date** ____________________

True-False

T F **1.** In the USDA Institutional Meat Purchase Specifications, cuts of beef are numbered by weight.

T F **2.** Beef sirloin can be cut into steaks and marinated to increase tenderness and then grilled or broiled.

T F **3.** T-bone steaks include the smaller section of the tenderloin.

T F **4.** A beef shank is a primal cut of beef that includes a thin portion of the beef forequarter located just beneath the rib cut.

T F **5.** To determine the doneness of whole tongue, the tip of the tongue is touched to test tenderness.

T F **6.** Brisket is a tender cut of beef that becomes tougher when cooked.

Multiple Choice

____________ **1.** The brisket and ___ are two separate muscle groups that make up one primal cut of beef.
A. round
B. sirloin
C. flank
D. shank

____________ **2.** ___ are a convenient way of providing uniform portions while reducing labor costs.
A. Whole carcasses
B. Partial carcasses
C. Fabricated cuts
D. Primal cuts

____________ **3.** A side of beef that is cut between the ___ ribs yields two quarters of beef.
A. 8th and 9th
B. 12th and 13th
C. 15th and 16th
D. 18th and 20th

_______________ **4.** ___ are considered the second most tender cut in the carcass and are ideal for grilling or broiling.

A. Flat-iron steaks
B. Shoulder petite tender roasts
C. Chuck rolls
D. Chuck eye steaks

_______________ **5.** A beef ___ is a primal cut of beef that includes the thin, flat section of the hindquarters located beneath the loin.

A. flank
B. round
C. sirloin
D. short plate

_______________ **6.** ___ cuts include ground beef, stew meat, kebab meat, and beef strips.

A. Subprimal
B. Primal
C. Prepared
D. Fabricated

Completion

_______________ **1.** The ___ is the largest primal cut, and its average weight is approximately 26% of the total carcass weight.

_______________ **2.** Except for ___, beef variety meats are prepared using moist-heat cooking methods.

_______________ **3.** Tripe is the muscular inner lining of a(n) ___ of an animal, such as cattle or sheep.

_______________ **4.** A side of beef is a(n) ___ of a carcass split along the backbone.

_______________ **5.** A beef strip loin is a short loin without a(n) ___.

_______________ **6.** A rib eye is a large, eye-shaped muscle within the rib that is a continuation of the ___ muscle.

Chapter 18 Review
Beef, Veal, and Bison

Section 18.3 Veal

Name ______________________ Date ______________

True-False

T F **1.** Veal has more connective tissue and fat than beef.

T F **2.** Nearly all veal calves that are slaughtered are male.

T F **3.** Special-fed veal calves are commonly raised in barns and harvested between 18 and 22 weeks of age.

T F **4.** Pasture-raised veal is less sustainable than other types of veal.

Multiple Choice

______________ **1.** Most veal calves are slaughtered at ___ months of age or younger.
A. three
B. four
C. five
D. six

______________ **2.** ___ veal is the flesh of a male dairy calf that is harvested at up to three weeks of age.
A. Special-fed
B. Bob
C. Pasture-raised
D. Grain-fed

______________ **3.** ___ veal is the flesh of calves that are not confined and are raised on a diet that consists of grass and their mothers' milk.
A. Bob
B. Special-fed
C. Pasture-raised
D. Grain-fed

______________ **4.** Raw veal that is extremely pale indicates a lack of ___ in the calf's diet.
A. vitamin B
B. protein
C. calcium
D. iron

Completion

__________________ **1.** ___ is the flesh of calves, which are young cattle.

__________________ **2.** Grain-fed veal calves are usually between the ages of six and ___ months when brought to market.

__________________ **3.** Pasture-raised veal calves have flesh that is close to ___ in color with a distinct flavor profile.

__________________ **4.** ___ veal, also referred to as milk-fed or formula-fed veal, is the flesh of calves that are fed an all-liquid milk-replacer diet that includes essential nutrients.

Name ______________________________ **Date** ________________

True-False

T F **1.** Veal leg is the least versatile cut of veal because it contains solid, lean, fine-textured meat.

T F **2.** The primal cuts of veal include the shoulder, rack, loin, leg, and foreshank and breast.

T F **3.** Veal is typically not split into sides like beef.

T F **4.** A veal rack accounts for approximately 25% of the total carcass weight.

Multiple Choice

______________ **1.** A veal ___ is a primal cut that contains the first four rib bones, some of the backbone, a small amount of each arm, and the blade bones.
A. shoulder
B. rack
C. loin
D. foreshank

______________ **2.** Veal liver is commonly broiled, pan-fried, or sautéed and should be cooked ___.
A. rare
B. medium-rare
C. medium
D. well-done

______________ **3.** A veal hindsaddle is the rear half of a carcass that includes the ___ and leg.
A. shoulder
B. rack
C. shank
D. loin

______________ **4.** A veal ___ can be split into halves and tied into a circle to form a crown rib roast.
A. leg
B. rack
C. loin
D. round

Completion

______________________ **1.** ___ are the thymus glands located in the neck of a calf.

______________________ **2.** A veal ___ is a thin slice of veal.

______________________ **3.** A(n) ___ is a small, ¼-inch thick veal cutlet that is generally sliced from leg meat and is commonly 2–3 inches in diameter.

______________________ **4.** The veal rack contains ___ rib bones on each side.

SECTION 18.5 INSPECTION AND GRADES OF BEEF AND VEAL

Name ____ **Date** ____________________

True-False

T F **1.** USDA Prime is the least flavorful of all meats.

T F **2.** Veal can only be graded for quality.

T F **3.** At the time of slaughter, beef and veal carcasses are stamped to indicate the quality of the meat that will be cut from the carcass.

T F **4.** USDA quality grading of beef and veal is voluntary.

Multiple Choice

________________ **1.** Most foodservice operations prefer USDA ___ beef because it is economical and the meat is tender, juicy, and flavorful.
A. Prime
B. Choice
C. Select
D. Good

________________ **2.** The ___ on the USDA inspection stamp identifies the plant where the animal was processed.
A. abbreviation
B. symbol
C. letter
D. number

________________ **3.** ___ grades are used to indicate tenderness, juiciness, and flavor and are based on specific criteria, such as the degree of marbling in the meat.
A. Quality
B. Yield
C. Processing
D. Identification

________________ **4.** USDA ___ is not a grade used with veal.
A. Prime
B. Choice
C. Select
D. Good

Completion

____________________ **1.** Beef and ___ are the only animals graded for yield.

____________________ **2.** USDA ___ beef is lean and has minimal marbling.

____________________ **3.** Most high-quality beef will have a(n) ___ grade of 1, 2, or 3.

____________________ **4.** USDA ___ beef and veal have abundant marbling.

SECTION 18.6 RECEIVING AND STORING BEEF AND VEAL

Name ______________________________ **Date** ______________

True-False

T F **1.** Vacuum-packed beef and veal should not be opened until needed for service or preparation.

T F **2.** Beef and veal are potentially hazardous foods that must be checked for color, odor, texture, and flavor upon receipt.

T F **3.** When frozen meats need to be thawed, they should be placed in the refrigerator overnight.

Multiple Choice

______________ **1.** Veal should be ___ with white fat.
A. white
B. pink
C. red
D. grey

______________ **2.** Veal should have no odor, and the texture should be ___.
A. firm
B. soft
C. dry
D. slick

______________ **3.** Once the vacuum seal is broken, vacuum-packed meat has a shelf life of only ___.
A. two to three days
B. four to five days
C. one to two weeks
D. two to three weeks

Completion

______________ **1.** Refrigerated beef and veal should maintain an internal temperature of ___°F or below.

______________ **2.** Beef or veal that is in the ___ should be rejected.

______________ **3.** Vacuum-packed beef and veal can be stored refrigerated for three to four ___.

Chapter 18 Review
Beef, Veal, and Bison

Section 18.7 Fabrication of Beef and Veal

Name ______________________________ **Date** ____________________

True-False

T F **1.** To cut a boneless strip loin into steaks, the loin is cut with the grain into steaks of desired weight or thickness.

T F **2.** One way to tenderize meat is by slicing it across the grain to produce thin slices with shorter muscle fibers.

T F **3.** Meat that is to be ground should be room temperature before grinding.

Multiple Choice

__________ **1.** To trim and cut beef tenderloin, a stiff boning knife is first used to carefully remove the ___ muscle from the side of the tenderloin.
A. adductor
B. main
C. wing tip
D. chain

__________ **2.** Beef tenderloin is often trimmed and then cut into portion-controlled cuts such as tenderloin tips, chateaubriand, ___, and tournedos.
A. filets mignons
B. prime ribs
C. strip loins
D. T-bone steaks

__________ **3.** Bone-in veal chops are commonly ___ for service.
A. sliced
B. Frenched
C. pounded
D. butterflied

Completion

__________ **1.** To begin Frenching veal chops, the ___ is peeled back to where it is still connected to the rack and then removed completely.

____________________ **2.** ___ beef and veal roasts holds the meat in a consistent shape that promotes even cooking.

____________________ **3.** The ___ end of the strip loin contains connective tissue that does not break down during cooking.

Chapter 18 Review
Beef, Veal, and Bison

Section 18.8 Flavor Enhancers for Beef and Veal

Name ______________________________ **Date** ____________________

True-False

T F **1.** Summer sausage, beef jerky, and bresaola are common cured-beef products.

T F **2.** Meat that is experiencing rigor mortis is called "green meat" and should not be eaten because it is extremely tough and almost flavorless.

T F **3.** After dry aging meat, the dry surface and the mold are left on during cooking.

T F **4.** Dry rubs are made by grinding herbs and spices together into a fine powder.

T F **5.** Veal is typically aged for seven to ten days.

Multiple Choice

______________ **1.** ___ add flavor to meat and tenderize it at the same time.
A. Marinades
B. Herbs
C. Spices
D. Gravies

______________ **2.** While fatback prevents meat from drying out, it also prevents the meat from ___.
A. tenderizing
B. reaching a safe temperature
C. browning
D. aging

______________ **3.** ___ is the period of rest that occurs after an animal has been slaughtered.
A. Maturing
B. Larding
C. Aging
D. Curing

______________ **4.** The length of time that meat should be left in a marinade is determined by the cut and ___ of the piece of meat.
A. shape
B. size
C. age
D. texture

______________ **5.** Meats are considered dry aged after an average of ___.
A. 7–10 hours
B. 3–6 days
C. 7–15 days
D. 2–3 weeks

Completion

______________ **1.** Larded meats are commonly ___ to allow ample time for the fat to be rendered and absorbed by the meat.

______________ **2.** Beef is wet aged under normal refrigeration for an average of 7–15 ___.

______________ **3.** ___ is air-dried cured beef that is aged two to three months until it becomes hard and turns almost purple in color.

______________ **4.** The ___ liquid base of a marinade tenderizes the meat by breaking down the protein structure of the meat.

______________ **5.** Within 5–24 hours after slaughter, a carcass goes through a period known as ___ where natural enzymes in the meat cause the tissues to seize and become stiff.

Matching

______________ **1.** Wet aging

______________ **2.** Larding

______________ **3.** Dry aging

______________ **4.** Barding

A. The process of aging meat in vacuum-sealed plastic

B. The process of laying a piece of fatback across the surface of a lean cut of meat to add moisture and flavor

C. The process of aging larger cuts of meat that are hung in a well-controlled environment

D. The process of inserting thin strips of fatback into lean meat with a larding needle

Name ______________________________ Date ______________

True-False

T F **1.** Ground beef and veal that are done have been cooked to an internal temperature of 145°F.

T F **2.** Beef brisket is commonly smoked.

T F **3.** It is not important to allow for carryover cooking when roasting meats.

T F **4.** If meat is cut in the same direction as the grain, the result is typically a tough, chewy piece of meat.

T F **5.** Veal is used to make white stews such as a fricassee and a blanquette.

Multiple Choice

______________ **1.** ___ and stewing are combination cooking methods that can be used to prepare tough or tender cuts of beef and veal.

A. Smoking
B. Broiling
C. Simmering
D. Braising

______________ **2.** Lean cuts of meat will become dry when roasted unless some form of ___ is added.

A. spice
B. fat
C. acid
D. dry rub

______________ **3.** The most accurate way to determine the degree of doneness of meat, with the exception of braised or stewed meats, is by ___ of the meat.

A. smelling
B. touching or tasting
C. measuring the internal temperature
D. the resting appearance

______________ **4.** Thick beef and veal chops are initially pan-fried and then finished in a(n) ___.
A. wok
B. steamer
C. fryer
D. oven

______________ **5.** ___ the meat allows juices that seized up during cooking to redistribute throughout the meat.
A. Resting
B. Marinading
C. Dressing
D. Slicing

Completion

______________ **1.** ___ cuts of beef and veal can be cooked using dry-heat cooking methods.

______________ **2.** ___ is the loss of volume and weight of a piece of food as the food cooks.

______________ **3.** Simmering makes meat more tender by breaking down ___ tissue.

______________ **4.** Cooking meat at too ___ a temperature can toughen the protein.

______________ **5.** A bone-in rib roast, often referred to as ___, is a tender cut of meat that is often carved before service with or without the bone.

Chapter 18 Review
Beef, Veal, and Bison

Section 18.10 Bison

Name ______________________________ **Date** ____________________

True-False

T F **1.** The similarities to beef allow bison to be used in any recipe that calls for beef.

T F **2.** Bison is graded for quality and yield.

T F **3.** Bison has more marbling than beef.

Multiple Choice

______________ **1.** Most bison are raised free-range and ___.
- A. grass-fed
- B. grain-fed
- C. milk-fed
- D. special-fed

______________ **2.** Bison is typically lower in ___ and cholesterol than beef.
- A. protein
- B. saturated fat
- C. unsaturated fat
- D. quality

______________ **3.** Bison must be procured from ___-inspected plants.
- A. EPA
- B. CDC
- C. USDA
- D. FDA

Completion

______________ **1.** Bison meat is similar in texture and flavor to beef, but slightly richer and ___.

______________ **2.** A bison is a large animal similar to cattle that is over 6 feet in height and 10 feet in length and can weigh over ___ lb.

______________ **3.** Bison is sold in market forms comparable to those of ___.

Name ______________________________ Date ______________________

Activity: Identifying Primal Cuts of Beef and Veal

Beef and veal are available in a variety of market forms including whole and partial carcasses. These are larger cuts and adequate facilities and equipment for fabrication and storage are necessary. A primal cut is a division of a partial carcass cut.

Match each primal cut to its image.

______________ **1.** Veal shoulder

______________ **2.** Veal breast

______________ **3.** Beef chuck

______________ **4.** Beef shank

______________ **5.** Veal leg

______________ **6.** Beef short plate

______________ **7.** Beef short loin

______________ **8.** Veal loin

______________ **9.** Beef flank

______________ **10.** Veal foreshank

______________ **11.** Beef round

______________ **12.** Veal rack

______________ **13.** Beef rib

______________ **14.** Beef sirloin

______________ **15.** Beef brisket

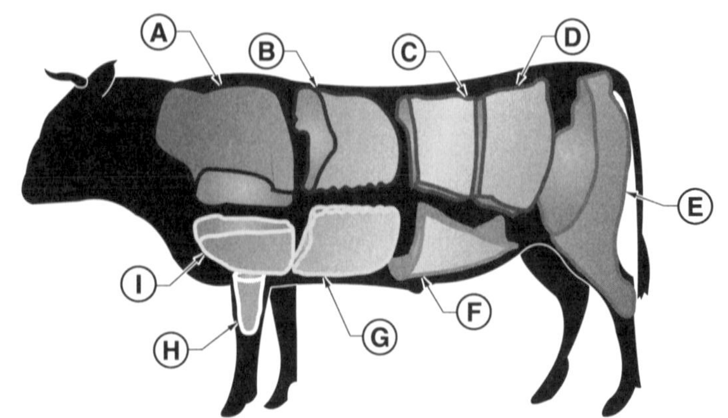

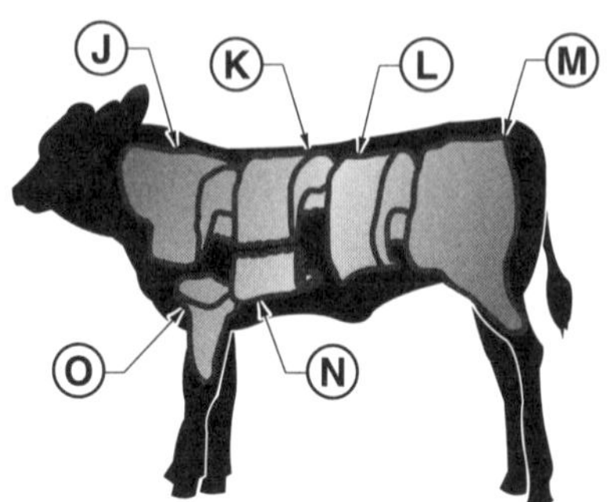

Activity: Identifying Fabricated Cuts of Beef

Each primal cut of beef is further divided into fabricated cuts. It is important to know the fabricated cuts to order product effectively and apply cooking methods that will result in a superior dish.

Meach of the following fabricated beef cuts to its image.

_______________ 1. Chuck roll

_______________ 2. Flat-iron steaks

_______________ 3. Rib roast

_______________ 4. Bone-in rib eye steak

_______________ 5. Tenderloin

_______________ 6. T-bone steak

_______________ 7. Filet mignons

_______________ 8. Bottom sirloin tri-tip

_______________ 9. Top sirloin butt

_______________ 10. Top round

_______________ 11. Eye of round

_______________ 12. Flank

_______________ 13. Short ribs

_______________ 14. Brisket

Photo Courtesy of The Beef Checkoff
(A)

Photo Courtesy of The Beef Checkoff
(B)

Photo Courtesy of The Beef Checkoff
(C)

Photo Courtesy of The Beef Checkoff
(D)

Photo Courtesy of The Beef Checkoff
(E)

Photo Courtesy of The Beef Checkoff
(F)

Photo Courtesy of The Beef Checkoff
(G)

Photo Courtesy of The Beef Checkoff
(H)

Photo Courtesy of The Beef Checkoff
(I)

Photo Courtesy of The Beef Checkoff
(J)

Photo Courtesy of The Beef Checkoff
(K)

Photo Courtesy of The Beef Checkoff
(L)

Canada Beef, Inc.
(M)

Photo Courtesy of The Beef Checkoff
(N)

Activity: Identifying Fabricated Cuts of Veal

Veal is available in a variety of market forms, including whole and partial carcasses, primal cuts, fabricated cuts, and variety meats. Knowledge of fabricated cuts is necessary to prepare accurate dishes and order effectively.

Match each of the following fabricated veal cuts to its image.

_______________ **1.** Bone-in shoulder

_______________ **2.** Stew meat

_______________ **3.** Trimmed rack

_______________ **4.** Chop

_______________ **5.** Frenched chop

_______________ **6.** Trimmed loin

_______________ **7.** Loin chops

_______________ **8.** Ossobuco

_______________ **9.** Scallopinis

_______________ **10.** Foreshank

_______________ **11.** Leg

Strauss Free Raised
(A)

Strauss Free Raised
(B)

Strauss Free Raised
(C)

Strauss Free Raised
(D)

Strauss Free Raised
(E)

Strauss Free Raised
(F)

Strauss Free Raised
(G)

Strauss Free Raised
(H)

Strauss Free Raised
(I)

Strauss Free Raised
(J)

Strauss Free Raised
(K)

Activity: Fabricating Beef and Veal

The size of storage facilities and availability of staff with fabrication skills often determine whether a food-service operation purchases primal cuts, fabricated cuts, or a combination. Beef and veal fabrication require trimming, cutting, tenderizing, and tying.

Perform the following procedures. Have an instructor check the result and sign below when complete.

1. Trim and cut a beef tenderloin.

Trimming and Cutting Beef Tenderloin Proficiency

Instructor:______________________________ Date:____________

2. Cut a boneless strip loin into steaks.

Cutting Boneless Strip Loin into Steaks Proficiency

Instructor:______________________________ Date:____________

3. French veal chops.

Frenching Veal Chops Proficiency

Instructor:______________________________ Date:____________

4. Tenderize a cut of beef with a mallet.

Tenderizing with a Mallet Proficiency

Instructor:______________________________ Date:____________

5. Tie a beef or veal roast.

Tying a Roast Proficiency

Instructor:______________________________ Date:____________

Activity: Preparing Beef

The manner in which beef is prepared for service can impact the tenderness of the final product. Seasonings, marinades, and slicing procedures can help produce a tender, flavorful dish.

Grill an unseasoned flank steak until it reaches an internal temperature of 145°F. Allow the steak to rest 3 minutes and slice it along the grain. Respond to the following.

1. Describe the texture and flavor of the unseasoned flank steak sliced along the grain.

Season and grill a flank steak until it reaches an internal temperature of 145°F. Allow the steak to rest 3 minutes and slice it across the grain. Respond to the following.

2. Describe the texture and flavor of the seasoned flank steak sliced across the grain.

Marinate and grill a flank steak until it reaches an internal temperature of 145°F. Allow the steak to rest 3 minutes and slice it across the grain. Respond to the following.

3. Describe the texture and flavor of the marinated flank steak sliced across the grain.

4. Which flank steak preparation was the most tender?

5. Which flank steak preparation was the most flavorful?

Activity: Determining Degrees of Doneness

Beef and veal cuts that are done have been cooked to an internal temperature of 145°F and then rested for 3 minutes. Ground beef and veal must be cooked to an internal temperature of 160°F. Beef and veal may be served to varying degrees of doneness.

Complete the following questions.

1. What is the internal temperature for very rare?

2. What is the internal temperature for rare?

3. What is the internal temperature for medium-rare?

4. What is the internal temperature for medium?

5. What is the internal temperature for medium-well?

6. What is the internal temperature for well-done?

Chapter 19 Review
Pork and Related Game

Section 19.1 Pork

Name ______________________________ **Date** ____________________

True-False

T F **1.** Pork is the meat from slaughtered hogs that are over a year old.

T F **2.** Pork muscles contain a large amount of fat.

T F **3.** Weight-bearing muscles and muscles that are continually used have the most connective tissue.

T F **4.** Hogs have more connective tissue in the legs and shoulders because these muscles are used frequently.

Multiple Choice

______________ **1.** ___ is trimmed from meat prior to cooking, as it is inedible.
- A. Collagen
- B. Elastin
- C. Marbling
- D. Muscle

______________ **2.** A(n) ___ is the fat that surrounds the outside of a muscle on a cut of meat.
- A. collagen
- B. elastin
- C. fat cap
- D. silverskin

______________ **3.** Cuts with a lot of connective tissue are best cooked using cooking methods such as ___.
- A. stewing
- B. grilling
- C. stir-frying
- D. broiling

______________ **4.** Connective tissue ___.
- A. yields tender cuts of meat
- B. yields flavorless cuts of meat
- C. breaks down easily
- D. does not break down easily

Completion

______________ **1.** Bones are often not removed from many cuts of pork prior to cooking in order to enhance ___ and presentation.

______________ **2.** Collagen is a soft, white connective tissue that breaks down into ___ when heated.

______________ **3.** ___ is a tough, rubbery, silver-white connective tissue that does not break down when heated.

______________ **4.** Leaving the fat cap on a piece of meat, such as chops, during cooking prevents the meat from ___.

SECTION 19.2 PURCHASING, RECEIVING, AND STORING PORK

Name ______________________________ **Date** ____________________

True-False

T F **1.** Back ribs are the meaty bones on the rib end of the pork belly.

T F **2.** Variety meats include parts of the hog such as the liver, kidneys, heart, tongue, jowls, hocks, feet, and ears.

T F **3.** Purchasing a whole hog carcass limits a chef's creativity with the menu.

T F **4.** Pork is available in a variety of market forms, including whole carcasses, primal cuts, fabricated cuts, and variety meats.

T F **5.** The five primal cuts of pork are the picnic shoulder, shoulder butt, loin, leg, and belly/spareribs.

T F **6.** A picnic is fabricated from the upper part of the hind leg.

T F **7.** Fresh ham is commonly cut from the middle of the shank bone to the aitchbone, or hip bone.

T F **8.** The USDA publishes the Institutional Meat Purchase Specifications (IMPS) for commonly purchased meats and meat products, including pork.

Multiple Choice

______________ **1.** Canadian bacon is the trimmed, pressed, and smoked boneless ___ of pork.

A. loin
B. belly
C. leg
D. shoulder

______________ **2.** A ___ cut is a ready-to-cook cut that is packaged to certain size and weight specifications.

A. primal
B. fabricated
C. partial
D. whole

_______________ **3.** The ___ contains the blade bone and a large portion of lean meat.
A. belly
B. loin
C. picnic shoulder
D. shoulder butt

_______________ **4.** Hocks are cut from the lower part of the ___ of a hog.
A. front and hind legs
B. front and hind feet
C. left and right shoulders
D. left and right ribs

_______________ **5.** Prosciutto is a type of ___-cured Italian ham that is sliced very thin and used to make hors d'oeuvres or appetizers.
A. dry
B. wet
C. combination
D. pump

_______________ **6.** Vacuum-packed pork helps to preserve the meat for ___.
A. ten to fifteen days
B. three to four weeks
C. one to two months
D. four to six months

_______________ **7.** Frozen pork should be stored at temperatures below ___°F.
A. 0
B. 10
C. 32
D. 41

_______________ **8.** ___ is unsmoked pork belly that has been cured in salt and spices, such as nutmeg and pepper, and then dried for about three months.
A. Pepperoni
B. Chorizo
C. Pancetta
D. Canadian bacon

Completion

_______________ **1.** Spareribs, pork belly, and bacon are fabricated cuts from the ___ primal cut.

_______________ **2.** Variety meats, also known as ___, are the edible parts of an animal that are not part of a primal cut.

_______________ **3.** ___ is a side pork that has been cured and usually smoked.

_______________ **4.** ___ pigs are available in the 10–30 lb range and are two to six weeks old.

_______________ **5.** ___ is the most tender cut of pork and can be prepared using any cooking method.

_______________ **6.** Pork ___ are the long, narrow ribs and breastbone of a hog.

Chapter 19 Review
Pork and Related Game

Section 19.3 Fabrication of Pork

Name ______________________ **Date** ______________

True-False

T F **1.** Boneless pork roasts stay intact when cooked.

T F **2.** Elastin does not need to be trimmed from a tenderloin.

T F **3.** A butterflied pork chop can be opened to lie flat.

Multiple Choice

______ **1.** To trim a tenderloin, the ___ is first carefully removed from the side of the tenderloin and reserved.
- A. fat cap
- B. silverskin
- C. rib bone
- D. chain muscle

______ **2.** If a loin is being fabricated in-house, the ___ must first be removed from the loin.
- A. tenderloin
- B. loin chop
- C. backbone
- D. clear plate

______ **3.** When tying a boneless pork roast, tie a(n) ___ to secure the roast before cutting the extra twine.
- A. slipknot
- B. overhand knot
- C. square knot
- D. fisherman's knot

Completion

______ **1.** To make a butterfly cut, hold the knife ___ to the cutting board.

______ **2.** A boneless roast is often rolled and tied with ___.

______ **3.** The final step in removing the tenderloin from the pork loin is to cut along the ___ bone until the tenderloin separates from the loin.

Name ______________________________ **Date** ________________

True-False

T F **1.** Adding salt to meat adds flavor and prevents microbial growth.

T F **2.** Combination curing is the process of combining either dry curing or immersion curing with stitch pumping to reduce processing time.

T F **3.** More than two-thirds of all pork is cured.

T F **4.** Dry rubs are made by grinding herbs and spices together into a fine powder and then rubbing the mixture into the pork after it is cooked.

T F **5.** Sodium from a brine is not absorbed into the leaner parts of meat.

Multiple Choice

_______________ **1.** Dry curing is a curing method that involves the use of salt to ___ the protein in food.
A. hydrate
B. dehydrate
C. increase
D. destroy

_______________ **2.** With immersion, meat is wet cured by ___.
A. coating the meat in a dry rub and then rinsing it
B. injecting water into the meat
C. submerging it in a brine solution
D. occasionally basting the meat with brine

_______________ **3.** ___ is done in stages to maximize the absorption of the curing ingredients.
A. Stitch pumping
B. Immersion
C. Wet curing
D. Dry curing

_______________ **4.** Sweet pickle brines are only used at ___.
A. temperatures above room temperature
B. room temperature
C. refrigeration temperatures
D. freezing temperature

_______________ **5.** The dry-curing time takes approximately ___ day(s) per pound for large cuts of pork.
A. one
B. two
C. three
D. four

Completion

_______________ **1.** Marinades have an acidic liquid base that ___ the meat by breaking down the protein structure of the pork.

_______________ **2.** A(n) ___ is a salt solution that usually consists of 1 cup of salt per 1 gal. of water.

_______________ **3.** Wet curing is a curing method in which foods are processed with a(n) ___.

_______________ **4.** Pork is the only meat that can have all its ___ cuts cured.

_______________ **5.** ___ is the process of using salt and sodium nitrite alone or with flavorings or sugar to preserve a food item.

Chapter 19 Review
Pork and Related Game

Section 19.5 Cooking Pork

Name ______________________________ **Date** ____________________

True-False

T F **1.** Pork is smoked and barbequed less often than other types meat.

T F **2.** The only difference between braising tender cuts and tougher cuts is the amount of time the pork is cooked.

T F **3.** Pork that is done has been cooked to an internal temperature of 145°F and has rested for 3 minutes.

T F **4.** An instant-read thermometer is inserted into the thinnest part of large cuts of pork to take the internal temperature.

T F **5.** Pork is typically basted in barbeque sauce during the grilling or smoking process.

Multiple Choice

______________ **1.** A pork chop that is cooked ___ is firm and springs back immediately when gently pressed with the fingertip.
A. well-done
B. medium
C. medium-rare
D. rare

______________ **2.** Large or tough cuts of pork are typically braised or ___.
A. sautéed
B. grilled
C. stewed
D. broiled

______________ **3.** ___ cuts of pork can be cooked with good results using dry-heat cooking methods.
A. Tough
B. Tender
C. Chewy
D. All

______________ **4.** ___ must be cooked to an internal temperature of 160°F.
A. Gound pork
B. Pork chops
C. Pork roasts
D. Pork steaks

______________ **5.** ___ pork requires that the meat be completely covered with liquid and cooked slowly.

A. Sautéing
B. Roasting
C. Frying
D. Stewing

Completion

______________ **1.** ___ such as apple and cherry emit less smoke as they burn, infusing meat with a light flavor and color.

______________ **2.** ___ is the loss of volume and weight of a piece of food as the food cooks.

______________ **3.** Larger roasts require a longer cooking time and should be roasted at lower temperatures, between ___°F and 325°F, to prevent excess shrinkage.

______________ **4.** Sautéing uses a small amount of hot ___ to sear and cook pork.

______________ **5.** Pork is typically cut into ___ for stir-frying.

Name ______________________________ **Date** ____________________

True-False

T F 1. Wild boar must be purchased from ranches or farms and processed in USDA-inspected plants.

T F 2. Although some cuts of wild boar are the same shape as pork cuts, wild boar meat is a much deeper red.

T F 3. Wild boar meat is not as lean as pork.

Multiple Choice

______________ 1. Wild boar raised for consumption are fed a diet consisting of ___.
- A. fruits and nuts
- B. small fish and frogs
- C. insects and leaves
- D. small animals, such as rabbits and squirrels

______________ 2. A popular cut of wild boar is the ___.
- A. rib rack
- B. shoulder
- C. shank
- D. tenderloin

______________ 3. Boar tenderloin is often ___.
- A. sautéed or fried
- B. grilled or smoked
- C. braised or stewed
- D. baked or broiled

Completion

______________ 1. ___ is a game animal that is similar in bone structure and muscle composition to domesticated hogs.

______________ 2. Wild boar meat has a stronger flavor and contains less ___ than pork.

______________ 3. The tenderloin of wild boar runs the length of the back from the hip to the ___.

CHAPTER 19 ACTIVITIES
PORK AND RELATED GAME

Name ____________________ **Date** ____________________

Activity: Identifying Primal Cuts of Pork

A hog carcass has two sets of primal cuts. One set is on the left side and one set is on the right side.

Match each primal cut of pork to its image.

______________ **1.** Belly

______________ **2.** Shoulder butt

______________ **3.** Picnic shoulder

______________ **4.** Loin

______________ **5.** Leg

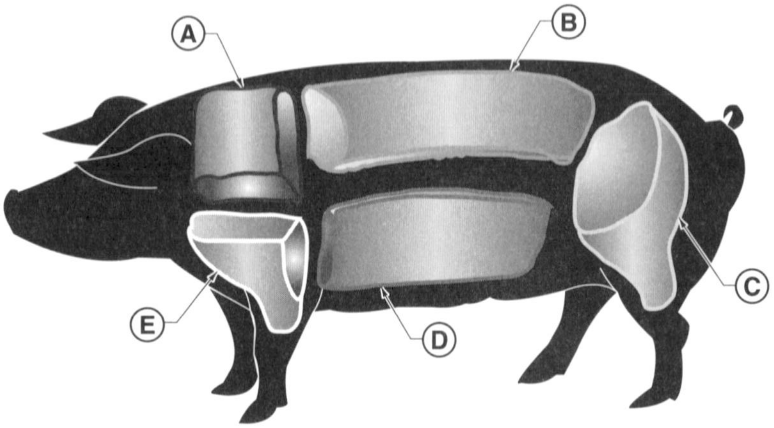

A

B

C

D

E

Activity: Identifying Fabricated Cuts of Pork

Each primal cut is divided into fabricated cuts. Fabricated cuts are a convenient way of providing uniform portions while reducing labor costs.

Match each of the following fabricated cuts of pork to its image.

_______________ **1.** Shoulder (Boston) butt

_______________ **2.** Blade steaks

_______________ **3.** Rib eye chops

_______________ **4.** Tenderloin

_______________ **5.** New York chops

_______________ **6.** Picnic (arm roast)

_______________ **7.** Leg (ham) primal cut

_______________ **8.** Cured ham

_______________ **9.** Spareribs

_______________ **10.** Bacon

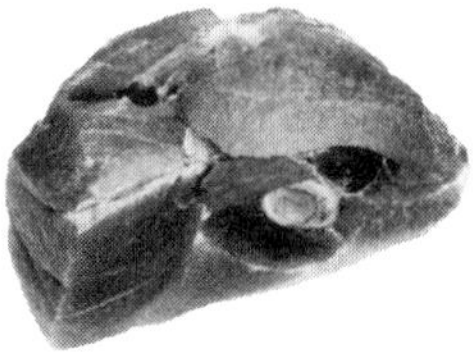

Courtesy of the National Pork Board

Ⓐ

Courtesy of the National Pork Board

Ⓑ

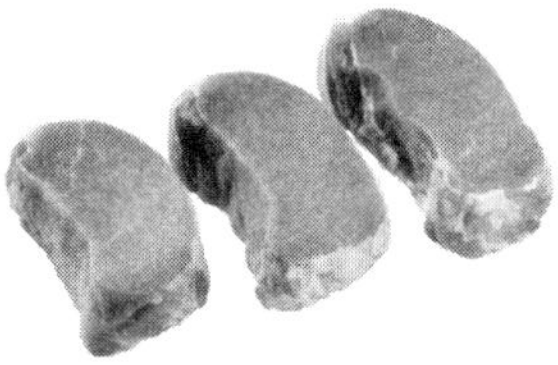

Courtesy of the National Pork Board

Ⓒ

Courtesy of the National Pork Board

Ⓓ

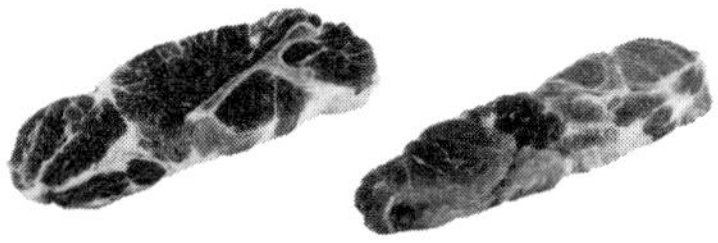

Courtesy of the National Pork Board

Ⓔ

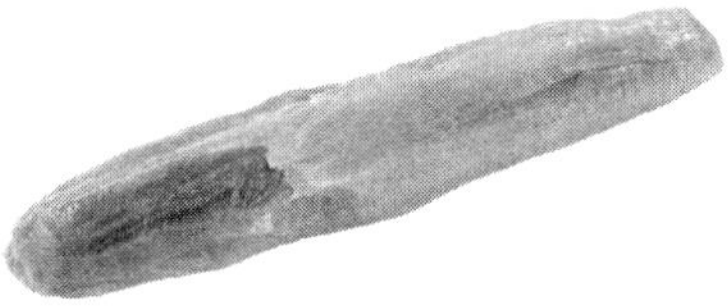

Courtesy of the National Pork Board

Ⓕ

Courtesy of the National Pork Board

Ⓖ

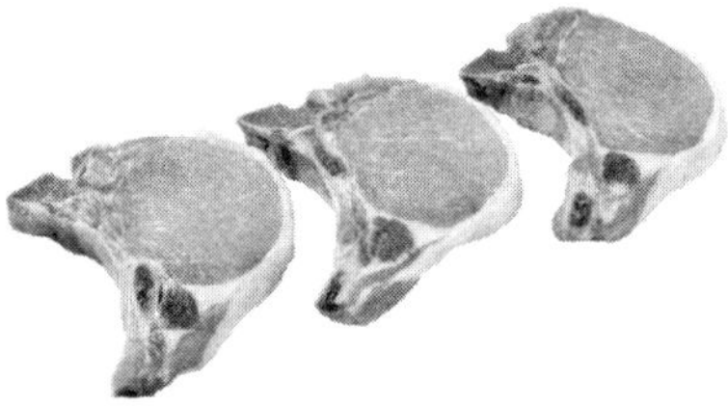

Courtesy of the National Pork Board

Ⓗ

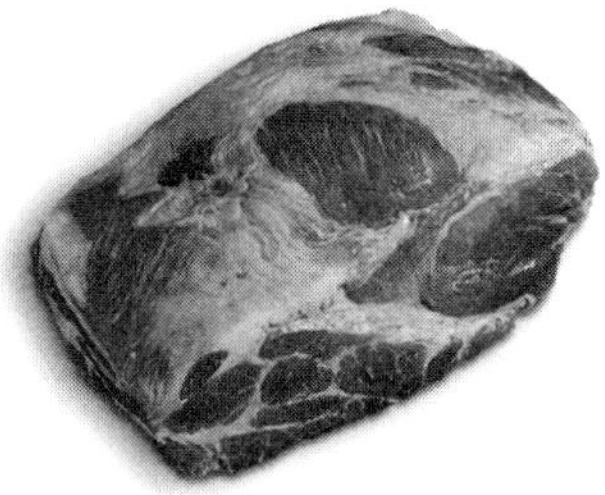

Courtesy of the National Pork Board

Ⓘ

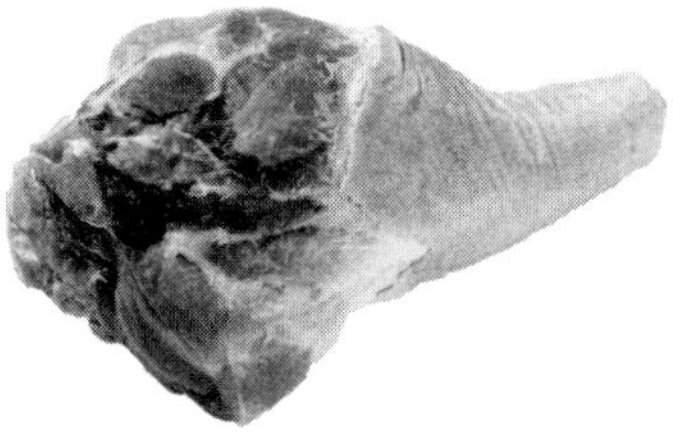

Courtesy of the National Pork Board

Ⓙ

Activity: Fabricating Pork

Some foodservice operations purchase whole hog carcasses and fabricate all the cuts in-house, while others only fabricate the loin. Pork fabrication requires the removal, trimming, and tying of meat as well as butterfly cuts.

Perform the following procedures. Have an instructor check the result and sign below when complete.

1. Remove the tenderloin from a pork loin.

Removing Tenderloin from a Pork Loin Proficiency

Instructor:____ Date:______________

2. Trim a pork tenderloin.

Trimming a Pork Tenderloin Proficiency

Instructor:____ Date:______________

3. Tie a boneless porkroast.

Tying a Boneless Pork Roast Proficiency

Instructor:____ Date:______________

4. Make a butterfly cut in a pork chop.

Butterfly Cut Proficiency

Instructor:____ Date:______________

Activity: Analyzing Curing Methods

Curing is the process of using salt and sodium nitrite alone or with flavorings or sugar to preserve a food item. Pork is the only meat that can have all its primal cuts cured.

Respond to the following.

1. What are the three methods used to cure pork?

2. Describe how meats are dry cured.

3. Why is salt used to dry cure meats?

4. Describe how meats are wet cured.

5. Compare the shelf life and flavor of a wet-cured product to a dry-cured product.

6. Describe the two ways to wet-cure meats.

7. What is combination curing?

Activity: Preparing Pork and Wild Boar

Pork and wild boar are similar in bone structure and muscle composition. All pork and wild boar served must be procured from USDA-inspected plants.

Examine a pork tenderloin and a boar tenderloin. Respond to the following.

1. Compare the appearance of the pork and boar tenderloin.

Season and grill both tenderloins until they each reach an internal temperature of 145°F and then let them rest for 3 minutes. Taste both tenderloins. Respond to the following.

2. Describe the appearance, texture, and flavor of the pork tenderloin.

3. Describe the appearance, texture, and flavor of the boar tenderloin.

4. Compare the appearance, texture, and flavor of the boar tenderloin to the pork tenderloin.

Chapter 20 Review
Lamb and Specialty Game

Section 20.1 Lamb

Name ______________________________ **Date** ____________________

True-False

T F **1.** Lamb is the meat from slaughtered sheep that are less than a year old.

T F **2.** Marbling is the fat that surrounds a muscle.

T F **3.** Elastin is not trimmed from meat prior to cooking because it makes the meat more tender.

T F **4.** Marbling can have an effect on the flavor, tenderness, and juiciness of lamb meat.

Multiple Choice

____________ **1.** Collagen is a soft, white, ___ that breaks down into gelatin when heated.
A. carbohydrate
B. protein
C. connective tissue
D. muscle fiber

____________ **2.** Lamb meat has a smooth grain and is similar in color to ___.
A. beef
B. chicken
C. veal
D. wild boar

____________ **3.** Muscles that receive ___ exercise have more connective tissue and are less tender.
A. no
B. slight
C. slight to moderate
D. the most

Completion

____________ **1.** Lamb ___ are porous and add flavor to the meat during the cooking process.

____________ **2.** ___ is the fat found within a muscle.

_______________ **3.** Elastin, also known as silverskin, is a tough, rubbery, silver-white ___ that does not break down when heated.

_______________ **4.** Lamb meat is made up of bundles of muscle fibers held together by two types of connective tissues called ___ and elastin.

Chapter 20 Review
Lamb and Specialty Game

Section 20.2 Purchasing, Receiving, and Storing Lamb

Name ______________________________ **Date** ______________________

True-False

T F **1.** A lamb rack is eight rib bones located between the shoulder and loin of a lamb.

T F **2.** The primal cuts of lamb are the shoulder, rack, loin, leg, and breast/shank.

T F **3.** Shoulder meat is most often cooked whole because of its many small bones and connective tissues.

T F **4.** Whole and partial carcasses of lamb are often not purchased due to the skilled labor and storage space required to process them.

T F **5.** An unsplit primal lamb loin is commonly known as a saddle.

T F **6.** Lamb legs are not split into two separate legs.

T F **7.** Quality grade shields are numbered 1 to 5 and indicate how much usable meat can be obtained from a lamb carcass.

T F **8.** USDA quality and yield grading is mandatory for lamb producers.

Multiple Choice

______________ **1.** An unsplit primal lamb rack is commonly known as a ___.
A. bracelet
B. rib roast
C. hotel rack
D. crown

______________ **2.** Meat from the shank end is most commonly used for ___ or ground for patties.
A. stew meat
B. variety meats
C. sweetbreads
D. filet mignon

______________ **3.** The ___ is the buttock, or rump bone, and is located at the top of the leg.
A. bracelet
B. chine bone
C. shank
D. aitchbone

______________ **4.** A lamb ___ is the upper section of the front leg located just below the shoulder and just above the knee.
A. foresaddle
B. foreshank
C. bracelet
D. riblet

______________ **5.** The foresaddle and hindsaddle are split ___.
A. down the backbone
B. below the ribcage
C. between the 6th and 7th ribs
D. between the 12th and 13th ribs

______________ **6.** USDA ___ is the most popular grade of lamb used in foodservice operations.
A. Prime
B. Choice
C. Select
D. Standard

______________ **7.** If necessary, lamb can be thawed under running water if it is cooked ___.
A. immediately after thawing
B. within 6 hours
C. using a dry-heat cooking method
D. using a slow cooking method

______________ **8.** Refrigerated lamb should maintain an internal temperature of ___°F or below.
A. 41
B. 52
C. 65
D. 87

Completion

______________ **1.** A lamb ___ consists of a rack and a loin that are still joined.

______________ **2.** A lamb ___ includes the first four rib bones of each side and the arm and neck bones.

______________ **3.** ___ is a method of removing the meat and fat from the end of a bone and is generally done to chops.

______________ **4.** A bracelet is a(n) ___ with the breast still attached.

______________ **5.** A(n) ___ of lamb includes part of the sirloin, the top round, bottom round, and knuckle meat.

______________ **6.** A(n) ___ cut is a ready-to-cook cut that is packaged to certain size and weight specifications.

________________ **7.** The lamb ___ is the rear half of the carcass that consists of the primal loin and primal leg.

________________ **8.** At the time of ___, the lamb carcass or the inspection tag is stamped with the round USDA inspection stamp, indicating the lamb was slaughtered at an inspected plant.

Matching

________________ **1.** Back

________________ **2.** Crown roast

________________ **3.** Lamb breast

________________ **4.** Partial carcass

________________ **5.** Frenching

________________ **6.** Riblet

________________ **7.** Noisette

________________ **8.** Bracelet

________________ **9.** Hindsaddle

________________ **10.** Lamb rack

________________ **11.** Lamb loin

________________ **12.** Foresaddle

A. Eight rib bones located between the shoulder and loin of a lamb

B. Method of removing the meat and fat from the end of a bone and is generally done to chops

C. Rack with the breast still attached

D. Frenched rack with the ribs formed into a circle to resemble a crown

E. Primal cut located between the rack and leg that includes the 13th rib, the loin eye muscle, the center section of the tenderloin, the strip loin, and some flank meat

F. Spareribs are commonly sliced between each rib bone to yield individual portions

G. Front half of a carcass that consists of the primal shoulder, primal rack, shank, and breast

H. Thin, flat cut of lamb that contains the breastbone, the sparerib bones, and the cartilage located under the shoulder and ribs

I. Rack and loin that are still joined

J. Small, round, and boneless medallion of meat

K. Rear half of a carcass that consists of the primal loin and primal leg

L. Primary division of a whole carcass

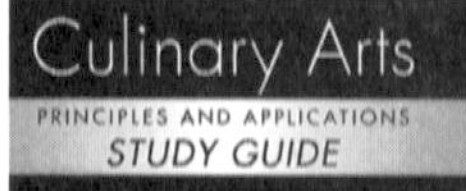

Chapter 20 Review
Lamb and Specialty Game

Section 20.3 Fabrication of Lamb

Name ______________________________ **Date** ____________________

True-False

T F **1.** The first step in fabricating a rack is to separate the hotel rack.

T F **2.** When cutting noisettes, the loin is cut lengthwise.

T F **3.** To start separating a hotel rack, turn the rack upside down, with the ribs pointing upward.

T F **4.** The hotel rack yields two separate racks.

T F **5.** A rack of lamb is never frenched.

Multiple Choice

______________ **1.** When Frenching a rack, the meat beneath the clean bones is referred to as the ___.
A. hotel rack
B. riblet
C. noisette
D. rib eye

______________ **2.** A square-cut whole shoulder is an economical lamb cut that is commonly fabricated into ___.
A. filets and steaks
B. rib racks and hotel racks
C. kebab and stew meats
D. bracelets

______________ **3.** The first step in fabricating a whole shoulder is to trim away the ___ and excess fat.
A. fell
B. marbling
C. aitchbone
D. chain muscle

______________ **4.** The ___ is a tough tendon that runs parallel to the vertabrae.
A. fell
B. back strap
C. silverskin
D. collagen

Completion

______________ **1.** ___ are small, round, and boneless medallions of meat.

______________ **2.** A rolled roast is ___ to maintain a consistent shape and ensure even cooking.

______________ **3.** ___ is a thin, tough membrane that lies directly under the hide and over the fat layer.

______________ **4.** A lamb ___ is commonly cut into noisettes.

Chapter 20 Review
Lamb and Specialty Game

Section 20.4 Flavor Enhancers for Lamb

Name ______________________ Date ______________

True-False

T F **1.** Marinades have an alkaline liquid base that tenderizes the meat by breaking down the protein structure of the lamb.

T F **2.** Very lean cuts of lamb can be barded.

T F **3.** After the meat has finished cooking, the fatback used for barding is removed.

Multiple Choice

______________ **1.** ___ is the process of laying a piece of fatback across the surface of a lean cut of meat to add moisture and flavor.
- A. Curing
- B. Marinating
- C. Larding
- D. Barding

______________ **2.** Common ___ for lamb include red wine, lemon juice, and yogurt.
- A. marinades
- B. dry rubs
- C. spice blends
- D. garnishes

______________ **3.** The length of time that lamb is left in the marinade depends on the ___ of the meat.
- A. temperature
- B. size and cut
- C. quality rating
- D. color

Completion

______________ **1.** ___ are made by grinding herbs and spices together into a fine powder and then rubbing the mixture onto the meat prior to cooking.

____________________ **2.** Once the meat has been removed from the marinade, the remaining liquid must be discarded to prevent ___.

____________________ **3.** In addition to slow, moist-heat cooking methods, certain ingredients and methods can help break down ___ and make the lamb meat more tender.

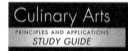

Chapter 20 Review
Lamb and Specialty Game

Section 20.5 Cooking Lamb

Name ______________________________ **Date** ____________________

True-False

T F **1.** Time is an accurate way to determine the doneness of meats.

T F **2.** Smaller loin roasts should be cooked at higher temperatures to allow them to caramelize on the exterior without overcooking the interior.

T F **3.** Only tougher cuts of lamb should be grilled.

T F **4.** Lamb cutlets and chops are typically pan-fried.

T F **5.** Searing the meat helps add flavor to the lamb and the resulting sauce.

Multiple Choice

______________ **1.** To determine the doneness of lamb, an instant-read thermometer is inserted into the ___ part of the meat to take the internal temperature.
A. thickest
B. thinnest
C. darkest
D. lightest

______________ **2.** The majority of animal muscle is made up of ___.
A. fat
B. collagen
C. protein
D. water

______________ **3.** Ground lamb must be cooked to an internal temperature of ___°F.
A. 145
B. 160
C. 175
D. 180

______________ **4.** When ___, the item is first browned in a small amount of fat, and then a cooking liquid is added to about halfway up the side of the meat.
A. roasting
B. sautéing
C. stewing
D. braising

______________ **5.** A lamb chop that is cooked medium-rare should have an internal temperature of ___°F.
A. 125
B. 135
C. 145
D. 155

Completion

______________ **1.** Sautéing uses a small amount of hot ___ to sear and cook the lamb.

______________ **2.** Tougher cuts of lamb can be slowly cooked using ___-heat cooking methods.

______________ **3.** A(n) ___ of lamb consists of pieces of lamb meat that are blanched and then rinsed to remove any impurities before being added to a cooking liquid.

______________ **4.** ___ is the loss of volume and weight of a piece of food as the food cooks.

______________ **5.** Carryover cooking affects the degree of doneness because the internal temperature of the meat continues to ___ after the meat is removed from the heat.

Name ____ **Date** __________________________

True-False

T F **1.** Goat meat is prepared in a similar manner to lamb meat.

T F **2.** Rabbit is similar to chicken in texture and can be used as a substitute in almost any chicken recipe.

T F **3.** Young bears, known as kids, have very sweet, tender meat.

T F **4.** Venison is a dark, lean meat with little to no marbling.

T F **5.** Uncooked rabbit meat looks similar to beef.

Multiple Choice

__________________ **1.** Kangaroo meat is high in ___.
- A. protein
- B. sodium
- C. cholesterol
- D. fat

__________________ **2.** ___ is considered specialty game.
- A. Bison
- B. Boar
- C. Goat
- D. Ostrich

__________________ **3.** The most desirable cuts of bear meat come from the loin and ___.
- A. neck
- B. shoulders
- C. front legs
- D. rear legs

__________________ **4.** ___ were one of the first domesticated animals and are still farmed for both their meat and milk.
- A. Rabbits
- B. Goats
- C. Bears
- D. Deer

____________________ **5.** ___ meat has a flavor profile that is a cross between beef and pheasant.
A. Kangaroo
B. Deer
C. Black bear
D. Goat

Completion

____________________ **1.** Game steaks and chops should reach an internal temperature of ___°F for 15 seconds.

____________________ **2.** ___ refers to the meat from deer, elk, antelope, moose, or pronghorn.

____________________ **3.** Bear meat should be cooked to the ___ stage.

____________________ **4.** Ranch-raised ___ bear is the most common type of bear meat served in restaurants.

____________________ **5.** Many quality cheeses are made from ___ milk.

Name ______________________ Date ______________

Activity: Identifying Primal Cuts of Lamb

Unlike beef, lamb is not split into sides before being divided into primal cuts. The left and right primal cuts remain joined together and are purchased as a single cut.

Match each primal cut of lamb to its image.

______________ **1.** Leg

______________ **2.** Loin

______________ **3.** Rack

______________ **4.** Shoulder

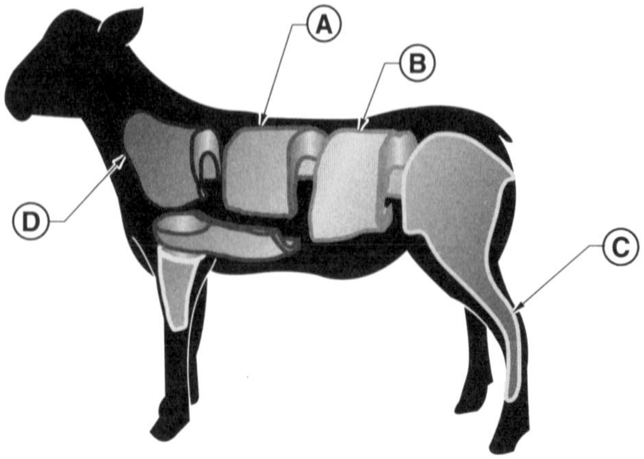

A

B

C

D

Activity: Identifying Fabricated Cuts of Lamb

The price per pound for fabricated cuts of lamb is higher than the price per pound for primal cuts. Most foodservice operations purchase some primal cuts and some fabricated cuts.

Match each of the following fabricated cuts of lamb to its image.

______ **1.** Square cut shoulder roast

______ **2.** Blade chop

______ **3.** Crown roast

______ **4.** Frenched rib chop

______ **5.** Boneless loin roast

______ **6.** Loin chop

______ **7.** Whole leg

______ **8.** Hindshank

______ **9.** Boneless sirloin roast

______ **10.** Foreshank

______ **11.** Riblets

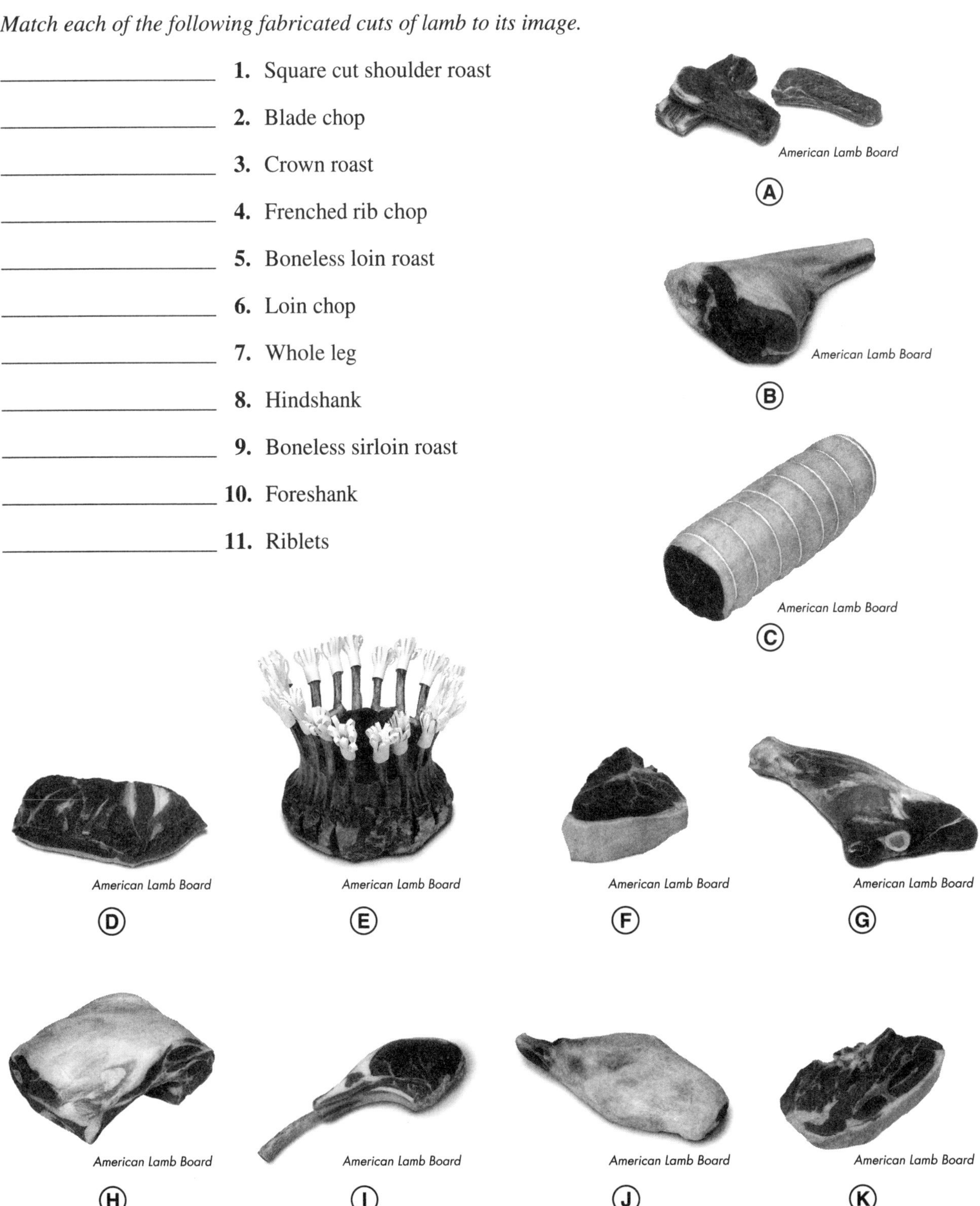

Activity: Fabricating Lamb

Lamb fabrication involves separating and frenching racks, boning and tying loins and legs, and cutting noisettes.

Perform the following procedures. Have an instructor inspect the result and sign below when complete.

1. Separate a hotel rack.

Separating a Hotel Rack Proficiency

Instructor:____ Date:_________________

2. French a rack of lamb.

Frenching a Rack of Lamb Proficiency

Instructor:____ Date:_________________

3. Bone a lamb loin to form a rolled roast.

Boning a Loin to Form a Rolled Roast Proficiency

Instructor:____ Date:_________________

4. Tie a rolled roast.

Tying a Rolled Roast Proficiency

Instructor:____ Date:_________________

5. Bone a leg of lamb.

Boning a Leg of Lamb Proficiency

Instructor:____ Date:_________________

6. Tie a leg of lamb.

Tying a Leg of Lamb Proficiency

Instructor:____ Date:________________

7. Cut lamb noisettes.

Cutting Noisettes Proficiency

Instructor:____ Date:________________

Activity: Analyzing Specialty Game

Venison, rabbit, goat, kangaroo, and bear are considered specialty game. These wild animals are raised on farms and ranches for their meat.

Match each of the following statements to the correct description.

_______________ **1.** Time and temperature requirement for game steaks and chops

_______________ **2.** Time and temperature requirement for game roasts

_______________ **3.** Time and temperature requirement for stuffed game meat

_______________ **4.** Time and temperature requirement for ground game meat

_______________ **5.** Game meat that tastes like a combination of lamb and beef

_______________ **6.** Game meat that when uncooked looks similar to skinless chicken meat

_______________ **7.** Game meat whose fat has a very strong, unpleasant taste and odor

_______________ **8.** Game meat that has a flavor that is a cross between beef and pheasant

_______________ **9.** Name referring to meat from deer, elk, antelope, moose, or pronghorn

A. Bear

B. 145°F for 4 minutes

C. Rabbit

D. Venison

E. 145°F for 15 seconds

F. Kangaroo

G. Goat

H. 165°F for 15 seconds

I. 160°F for 15 seconds

Activity: Researching Lamb and Specialty Game

The tender meat of lamb makes it ideal for creating a variety of dishes. Specialty game meats are growing in popularity.

Choose a lamb dish or a specialty game dish and write an essay using the following outline.

I. Introduction
II. History of the lamb or specialty game dish
  A. Country of origin
  B. Economic influences
III. Contemporary use
  A. Country/countries where the lamb or specialty game dish is commonly served
  B. Cultural influences
  C. Common variations
IV. Recipe
V. Conclusion

1. Prepare and present a report about a lamb or specialty game dish using the outline provided. Use visuals such as pictures and demonstrations to enhance the report. Include a list of the sources used for the research project.

Chapter 21 Review
Baking and Pastry Fundamentals

Section 21.1 Bakeshop Ingredients

Name ______________________________ Date ____________________

True-False

T F 1. Baking soda is an alkaline chemical leavening agent that reacts to an acidic dough or batter by releasing carbon dioxide with the addition of heat.

T F 2. When adding vanilla, the bean is ground and then added to the mixture.

T F 3. The proteins in milk assist in producing a coarser crumb in yeast breads.

T F 4. The most basic way to leaven baked products is to incorporate air and steam into a dough or batter.

T F 5. Milk used in the preparation of breads and sweet doughs may be liquid or dried.

T F 6. Ingredients cannot be easily substituted in baked products.

Multiple Choice

______________ **1.** Gluten is the protein in flour that, when combined with ___, gives a baked product its structure.
- A. water
- B. oil
- C. salt
- D. baking soda

______________ **2.** ___ adds flavor, color, and tenderness to baked products through caramelization.
- A. Flour
- B. Sugar
- C. Yeast
- D. Butter

______________ **3.** An extract is a flavorful oil that has been mixed with ___.
- A. alcohol
- B. water
- C. salt
- D. sugar

______________ **4.** ___ function as a thickener in custards and add moisture, leavening, and nutritive value to cake batters.
A. Purées
B. Sugars
C. Oils
D. Eggs

______________ **5.** ___ yeast is a form of yeast that does not need to be hydrated and may be directly added to a warm flour mixture.
A. Instant
B. Compressed
C. Active
D. Fresh

______________ **6.** ___ assist with moisture absorption and are often used to make icings and cakes that contain more sugar than flour.
A. Butters
B. Oils
C. Emulsified shortenings
D. Hydrogenated shortenings

Completion

______________ **1.** ___ flour is unbleached flour that still contains the bran, endosperm, and germ of the wheat kernel.

______________ **2.** Common ___ used in the bakeshop include cornstarch, arrowroot, flour, gelatin, tapioca, and modified starches.

______________ **3.** ___ adds richness and tenderness and improves grain, texture, and shelf life of baked products.

______________ **4.** ___ chocolate is pure chocolate liquor that contains 50% to 58% cocoa butter.

______________ **5.** ___ is a process that changes the molecular structure of an oil into a solid.

______________ **6.** ___ is a popular thickening agent because it will not cloud a clear sauce and is preferred for use in pie fillings.

______________ **7.** Dark brown sugar is a moist sugar product that contains approximately 7% ___.

______________ **8.** ___ is a fine powder that is created by grinding grains.

______________ **9.** ___ is a living, microscopic, single-celled fungus that releases carbon dioxide and alcohol through a process called fermentation when provided with food (sugar) in a warm, moist environment.

______________ **10.** Honey is approximately ___ times as sweet as granulated sugar.

Chapter 21 Review
Baking and Pastry Fundamentals

Section 21.2 Bakeshop Equipment

Name ______________________________ **Date** ____________________

True-False

T F **1.** Silicone mats may be used in the refrigerator or freezer but not in the oven.

T F **2.** Silicone bakeware is nonstick and temperature resistant.

T F **3.** A sheeter is an aluminum pan with low sides that is used to bake large amounts of cookies at one time.

T F **4.** Springform pans have a metal clamp on the side that allows the bottom of the pan to be separated from the sides.

Multiple Choice

______________ **1.** A ___ pan is a round, shallow baking pan with sloped sides that are smooth or fluted and may have a removable bottom.
- A. muffin
- B. pie
- C. tart
- D. cake

______________ **2.** A ___ is a dough-cutting tool with a rotating disk attached to a handle that is used to cut dough into desired shapes.
- A. dough cutter
- B. dough docker
- C. palette knife
- D. pastry wheel

______________ **3.** ___ are electronic appliances used to mix, blend, and whip ingredients.
- A. Mixers
- B. Sheeters
- C. Pastry wheels
- D. Revolving tray ovens

______________ **4.** ___ are used as a surface for rolling out dough and allow piecrust to be easily flipped from the cloth into pie pans.
- A. Silicon mats
- B. Pastry cloths
- C. Sheeters
- D. Wax paper squares

Completion

______________ **1.** A(n) ___ is a temperature- and humidity-controlled box that creates the perfect environment for yeast doughs to rise.

______________ **2.** ___ are aluminum or stainless steel rollers that have pins and are used to perforate dough so that it will bake evenly without blistering from the oven heat.

______________ **3.** A convection oven circulates ___ evenly while a product bakes.

______________ **4.** ___ knives are most often used to ice cakes.

Chapter 21 Review
Baking and Pastry Fundamentals

Section 21.3 Bakeshop Measurements, Formulas, and Percentages

Name ______________________________ **Date** ____________________

True-False

T F **1.** A baker's scale or a digital scale can be used to measure the weight of ingredients in the bakeshop.

T F **2.** In a bakeshop, ingredients are only measured by weight.

T F **3.** Because different ingredients have different densities, two items of the same volume can have different weights.

Multiple Choice

______________ **1.** A ___ is a recipe format in which precise measurements of each ingrendient are required to ensure successful results.

A. ratio
B. conversion
C. formula
D. baker's percentage

______________ **2.** Ingredient baker's percentages are calculated by dividing the ingredient weight by the main ingredient weight and multiplying that number by ___.

A. 10
B. 100
C. 10%
D. 100%

______________ **3.** The following formula is used to determine the ingredient weight for each ingredient: ___.

A. $IW = MW \times (BP \div 100)$
B. $MW = DY \div (T\% \div 100)$
C. $BP = (IW \div MW) \times 100$
D. $IW = BP \times (MW \div 100)$

Completion

______________ **1.** Unit measurements for ___ include grams (g), pounds (lb), and ounces (oz).

______________ **2.** A(n) ___ is the weight of an ingredient expressed as a percentage of the weight of the main ingredient (primary structure provider) in a formula.

______________ **3.** Each ingredient in a formula, including liquids, must be weighed using the same ___.

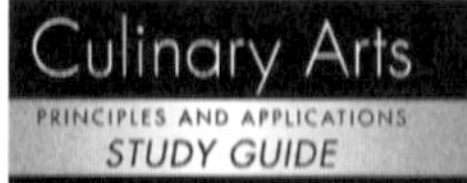

Name ______________________________ **Date** ______________

True-False

T F **1.** Rounding enables dough to proof evenly and have a smooth outer surface.

T F **2.** When scaling dough, the dough is torn by hand into pieces of a desired size before they are weighed.

T F **3.** Fermenting is the process of letting yeast dough rise in a warm and moist environment until the dough doubles in size.

T F **4.** A rich dough is a yeast dough that incorporates a significant amount of fat, sugar, and eggs into a heavy, soft structure.

T F **5.** A rolled-in dough is a yeast dough that has a flaky texture that results from the incorporation of fat through lamination.

Multiple Choice

______________ **1.** The ___ method is a method of mixing yeast dough in which the ingredients are added in sequential steps.
- A. straight-dough
- B. modified straight-dough
- C. sponge
- D. modified sponge

______________ **2.** Scoring is done on ___ breads before they are baked to allow carbon dioxide to escape during baking.
- A. rich-dough
- B. rolled-in
- C. hard-crusted
- D. laminated

______________ **3.** Fermentation will continue until dough reaches ___°F, the temperature at which yeast dies.
- A. 120
- B. 130
- C. 140
- D. 150

_______________ **4.** In yeast bread preparations, ___ acts as a stimulant to yeast.
A. sugar
B. milk or water
C. salt
D. fat

_______________ **5.** A ___ dough is a dough that is low in fat and sugar.
A. rich
B. dry
C. lean
D. rolled-in

Completion

_______________ **1.** A(n) ___ is a liquid that is brushed on the surface of a yeast dough product prior to baking.

_______________ **2.** ___ doughs are kneaded longer than other doughs because the high amounts of fat and sugar inhibit gluten development.

_______________ **3.** ___ is the rapid expansion of yeast dough in an oven, resulting from the expansion of gases within the dough.

_______________ **4.** Hard rolls, baguettes, and rye bread are made from ___ doughs.

_______________ **5.** ___ is the process of weighing an ingredient using a scale.

Matching

_______________ **1.** Panning

_______________ **2.** Kneading

_______________ **3.** Rounding

_______________ **4.** Proofing

_______________ **5.** Scoring

_______________ **6.** Docking

_______________ **7.** Fermentation

_______________ **8.** Lamination

A. The process of making shallow, angled cuts across the top of unbaked bread with a sharp knife called a lame

B. The process in which yeast dough undergoes its final rise after it has been shaped and before it is baked

C. The process of layering dough with butter so that it rises and expands as it bakes, resulting in a very flaky product

D. The process of shaping scaled dough into smooth balls

E. The process in which yeast releases carbon dioxide gas as it expands within a dough

F. The process of placing rounded pieces of dough into pans

G. The process of pushing and folding dough until it is smooth and elastic

H. The process of making small holes in yeast dough before it is baked to allow steam to escape and promote even baking

Chapter 21 Review
Baking and Pastry Fundamentals

Section 21.5 Quick Breads

Name ______________________________ **Date** ______________

True-False

T F **1.** A muffin is a quick bread made by mixing solid fat, baking powder or baking soda, salt, and milk with flour.

T F **2.** Muffins tend to be savory.

T F **3.** Solid fat, such as butter or shortening, helps produce a light and flaky product, such as a biscuit.

T F **4.** Quick breads may be frozen with excellent results.

T F **5.** When making corn breads, the liquid must be added slowly because cornmeal does not absorb liquid quickly.

Multiple Choice

______________ **1.** When baking quick breads, pans should be filled ___ full.
A. one-fourth
B. one-half
C. two-thirds
D. three-fourths

______________ **2.** ___ may be added to improve the tenderness and richness of biscuits.
A. Milk
B. Sugar
C. Butter
D. Salt

______________ **3.** ___ is a quick bread made from a batter containing cornmeal, eggs, milk, and oil.
A. Banana bread
B. Corn bread
C. Flatbread
D. Tortilla

______________ **4.** ___ are similar to muffins in quality and texture.
A. Pita breads
B. Corn breads
C. Biscuits
D. Quickbread loaves

______________ **5.** ___ supplies form and texture to biscuits.
A. Flour
B. Solid fat
C. Baking powder
D. Milk

Completion

______________ **1.** The ___ mixing method is a quick-bread mixing method that uses liquid fats, such as vegetable oil or melted butter, to produce a rich and tender product.

______________ **2.** A(n) ___ is a baked product that is made with a quick-acting leavening agent such as baking powder, baking soda, or steam.

______________ **3.** The creaming method creates a(n) ___ crumb than the muffin mixing method.

______________ **4.** If quick breads are not served immediately, they should be ___ after cooling.

______________ **5.** The creaming mixing method involves mixing ___ temperature fat and sugar in a mixer using the paddle attachment until the batter is light and fluffy.

Culinary Arts
PRINCIPLES AND APPLICATIONS
STUDY GUIDE

Chapter 21 Review
Baking and Pastry Fundamentals

Section 21.6 Cakes

Name ______________________________ **Date** ______________

True-False

T F **1.** The chiffon mixing method involves folding whipped egg whites into a batter made from flour, egg yolks, and fat.

T F **2.** During stage one of baking cakes, the lowest oven temperature called for in the baking instructions should be used.

T F **3.** Lighter cakes, such as sponge cakes, are tested for doneness by inserting a wire tester or a toothpick in the center of the cake.

T F **4.** The creaming method of mixing cake batter produces a dense product with a coarse crumb.

Multiple Choice

______________ **1.** In the ___ mixing method, eggs are added one at a time and incorporated into the creamed fat and sugar.

A. chiffon
B. creaming
C. sponge
D. two-stage

______________ **2.** The procedure for mixing cake batter using the ___ mixing method includes adding the egg mixture to the batter in thirds.

A. creaming
B. sponge
C. chiffon
D. two-stage

______________ **3.** The airy, whipped yolk and sugar mixture used in a genoise sponge mixing method is referred to as a ___.

A. foam
B. chiffon
C. meringue
D. cream

______________ **4.** During stage ___ of baking a cake, the top surface begins to brown.
A. 1
B. 2
C. 3
D. 4

Completion

______________ **1.** Whenever possible, cakes should be placed in the ___ of the oven where the heat is evenly distributed.

______________ **2.** Most cake recipes are developed to produce the best results when baked at or near ___.

______________ **3.** After the cake has cooled on the wire rack or shelf for a minimum of ___ minutes, the pan can be inverted and the cake removed.

______________ **4.** Although there are many variations of the sponge mixing method, the most common is referred to as a(n) ___.

CHAPTER 21 REVIEW
BAKING AND PASTRY FUNDAMENTALS

SECTION 21.7 ICINGS

Name ______________________________ **Date** ______________

True-False

T F **1.** When piping icing, the hand at the bottom of the pastry bag applies the pressure that causes the icing to flow, and the hand on the top half is the guide.

T F **2.** Foam icing is prepared by combining sugar, glucose, and water, boiling the mixture to approximately 240°F, and then adding the resulting syrup to an egg white meringue.

T F **3.** Typically, 1 cup of heavy cream yields 2 cups of whipped cream.

Multiple Choice

______________ **1.** A ___ icing is made by combining water, powdered sugar, corn syrup, and flavoring and then heating the mixture to 100°F.
- A. glaze
- B. flat
- C. fudge
- D. buttercream

______________ **2.** ___ is a rich, white icing that is prepared with heat and hardens when exposed to air.
- A. Ganache icing
- B. Foam icing
- C. Royal icing
- D. Fondant

______________ **3.** Glazes extend shelf life by ___.
- A. adding air into the icing
- B. coating perishable ingredients
- C. sealing in moisture
- D. adding preservatives

Completion

______________ **1.** Ganache icing is a rich chocolate icing made by pouring heated ___ over chocolate and stirring until the chocolate melts.

____________________ **2.** To form a paper pastry bag, a square of parchment paper is cut into a large ___.

____________________ **3.** A royal icing is similar to a flat icing, but the addition of ___ produces a thicker icing that hardens to a brittle texture.

Chapter 21 Review
Baking and Pastry Fundamentals

Section 21.8 Cookies

Name ______________________________ **Date** ____________________

True-False

T F **1.** Cookies are often named for their preparation method.

T F **2.** Like rolled cookies, molded cookies use a soft dough instead of a stiff dough.

T F **3.** Bar cookies are generally thicker than most types of cookies and often have a dense and cakelike texture.

Multiple Choice

__________ **1.** A soft cookie is a cookie prepared from dough that contains a high percentage of ___.
A. moisture
B. butter
C. sugar
D. flour

__________ **2.** ___ cookies are prepared from a moist, soft batter.
A. Molded
B. Refrigerator
C. Rolled
D. Drop

__________ **3.** To make rolled cookies, a fairly stiff dough is refrigerated until chilled and then rolled out to ___-inch thickness on a floured surface.
A. ⅛
B. ¼
C. ½
D. 1

Completion

__________ **1.** A(n) ___ cookie is a cookie prepared from dough that contains a high percentage of sugar.

__________ **2.** When making refrigerator cookies, the dough is wrapped and refrigerated until firm, and then the dough is cut into slices ___-inch thick.

__________ **3.** ___ cookies should be placed in an airtight tin container.

Chapter 21 Review
Baking and Pastry Fundamentals

Section 21.9 Pies

Name ______________________________ **Date** ______________________

True-False

T F **1.** French meringue, also known as common meringue, is made by whipping egg whites with a bit of lemon juice or cream of tartar.

T F **2.** Blind-baked crusts are often used for pies with unbaked fillings.

T F **3.** Meringues can be soft or hard, depending on the ratio of sugar to egg whites.

T F **4.** There are three types of flaky pie crusts: short-flake, medium-flake, and long-flake.

T F **5.** A double-crust pie consists of two bottom crusts.

Multiple Choice

________________ **1.** A ___ filling is an uncooked pie filling that is baked in an unbaked piecrust.
A. fruit
B. chiffon
C. cream
D. soft

________________ **2.** A ___ is a canned product such as fruit with little or no water added.
A. solid pack
B. soft filling
C. meringue
D. block pack

________________ **3.** A ___ crust is a piecrust made from crumbled cookies or crackers that are held together with melted butter.
A. flaky
B. mealy
C. crumb
D. specialty

________________ **4.** An Italian meringue is made by boiling sugar and water until it reaches ___°F, which produces a thick syrup.
A. 212
B. 225
C. 240
D. 265

______________ **5.** A ___ filling is a light, fluffy pie filling prepared by folding a meringue into a puréed fruit or a cream pie filling.
A. cream
B. chiffon
C. soft
D. fruit

Completion

______________ **1.** A(n) ___ piecrust is a piecrust prepared by cutting fat into flour until pea-size pieces of dough form.

______________ **2.** A(n) ___ is a mathematical way to represent the relationship between two or more numbers or quantities.

______________ **3.** ___ is a process in which a piecrust is baked for 10–15 minutes before a filling is added.

______________ **4.** A(n) ___ piecrust is a low-moisture piecrust prepared by rubbing fat into flour until the mixture resembles fine cornmeal.

______________ **5.** A(n) ___ is a mixture of egg whites and sugar.

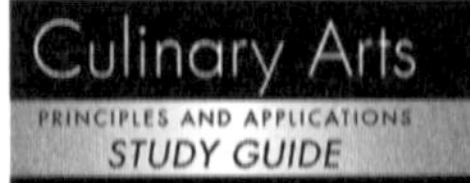

Name ______________________________ **Date** ____________________

True-False

T F **1.** Puff pastry is leavened with butter and yeast.

T F **2.** New York cheesecake is more dense and rich than Italian cheesecake.

T F **3.** Soufflés are not stable like cakes and easily collapse once removed from the oven.

T F **4.** A mousse contains a large amount of gelatin.

Multiple Choice

______________ **1.** ___ is a paper-thin dough in which each sheet of dough is brushed with butter and then layered.
- A. Pâte à choux
- B. Phyllo
- C. Puff pastry dough
- D. Pâte feuilletée

______________ **2.** ___ is a baked custard that has the texture of a stirred custard beneath a hardened sugar surface.
- A. Crème brûlée
- B. Crème caramel
- C. Cheesecake
- D. Flan

______________ **3.** Rice pudding is a baked custard made from cooked rice combined with a ___.
- A. savory custard
- B. condensed milk
- C. glaze icing
- D. sweet custard

______________ **4.** ___ is used as a base for other desserts, such as ice cream and Bavarian cream.
- A. Bread pudding
- B. Crème caramel
- C. Crème anglaise
- D. Crème brûlée

Completion

______________ **1.** A sabayon is prepared by whisking ___ and egg yolks over a double boiler until they are whipped into a silky foam.

______________ **2.** ___, also known as cream puff paste, is a pastry dough made by beating eggs into a paste of boiled water, butter, and flour.

______________ **3.** Crème caramel is also known as ___.

______________ **4.** Bavarian cream is a flavored custard sauce that is stabilized with ___ while it is still warm.

SECTION 21.11 FROZEN DESSERTS

Name ______________________________ **Date** ______________

True-False

T F **1.** Churning keeps an ice cream base aerated to prevent ice crystals from forming.

T F **2.** When three ice cream varieties are layered in a rectangular mold, the item is referred to as a bombe.

T F **3.** The use of milk instead of cream and eggs makes sherbet lower in fat than ice cream.

Multiple Choice

______________ **1.** Eggs add richness to ice cream and act as a(n) ___.
A. clarifying agent
B. emulsifier
C. leavening agent
D. flavoring

______________ **2.** The FDA allows a maximum of ___% overrun, although the best ice creams have less.
A. 25
B. 50
C. 75
D. 100

______________ **3.** A ___ has a grainy texture and ice crystals that are larger than those of a sorbet.
A. sherbet
B. bombe
C. granité
D. gelato

Completion

______________ **1.** ___ is a frozen dessert made from cream, butterfat, sugar, and sometimes eggs.

______________ **2.** ___ is an increase in the volume of a frozen product as a result of the incorporation of air during churning and freezing.

______________ **3.** A(n) ___ is an Italian ice cream that has a creamier, denser texture than standard ice creams.

Name ______________________________ **Date** ______________

Activity: Analyzing Bakeshop Ingredients

Different types of flours, sugars and sweeteners, fats, eggs, and milk produce different results when used in baked products. Thickeners and leavening agents are used to alter the texture and lightness of breads. The flavoring ingredient chosen often determines the flavor of the end product.

Answer the following questions.

1. What is the purpose of flour in bakeshop items?

2. What is the purpose of gluten in bakeshop items?

3. What types of flour are commonly used in bakeshop items?

4. What is the purpose of sugars and sweeteners in bakeshop items?

5. What types of sugars and sweeteners are commonly used in bakeshop items?

6. What is the purpose of pastry sugar in bakeshop items?

7. How is confectioners' sugar typically used in bakeshop items?

8. What is turbinado sugar?

9. What is the difference between dark brown sugar and light brown sugar?

10. What is molasses commonly added to in the bakeshop?

11. What does the flavor of honey depend upon?

12. What is the purpose of eggs in bakeshop items?

13. What is the purpose of fat in bakeshop items?

14. What types of fats are commonly used in bakeshop items?

15. What is European-style butter?

16. What is the purpose of emulsified shortenings in bakeshop items?

17. What is the purpose of milk in bakeshop items?

18. What is the purpose of thickening agents in bakeshop items?

19. What types of thickening agents are commonly used in bakeshop items?

20. What is the purpose of leavening agents in bakeshop items?

21. What types of leavening agents are commonly used in bakeshop items?

22. What types of flavorings are commonly used in bakeshop items?

Activity: Calculating Baker's Percentages

A baker's percentage is the weight of a particular ingredient expressed as a percentage based on the weight of the main ingredient in a formula. Formulas help to ensure that ingredients are measured consistently and accurately. To calculate the baker's percentage of an ingredient where flour is the main ingredient, apply the following formula:

$$BP = (IW \div MW) \times 100$$

where

BP = ingredient baker's percentage

IW = ingredient weight

MW = main ingredient weight

Calculate the baker's percentage for each ingredient listed. Show all calculations.

The recipe calls for 10 lb of flour.

____________________ **1.** The baker's percentage for 1 lb of sugar is ___%.

____________________ **2.** The baker's percentage for 2 lb of milk is ___%.

____________________ **3.** The baker's percentage for ½ lb of milk is ___%.

The recipe calls for 20 lb of flour.

____________________ **4.** The baker's percentage for 5 lb of sugar is ___%.

____________________ **5.** The baker's percentage for 1 lb of eggs is ___%.

____________________ **6.** The baker's percentage for ½ lb of butter is ___%.

The recipe calls for 5 lb of flour.

____________________ **7.** The baker's percentage for 2 lb of sugar is ___%.

____________________ **8.** The baker's percentage for 1 lb of chopped apples is ___%.

____________________ **9.** The baker's percentage for 3 lb of butter is ___%.

Activity: Weighing Dry Ingredients

Some recipes call for sifted flour while others do not. If a recipe calls for sifted flour, it is important to sift the flour before measuring the amount required. Even packages of flour marked "sifted" should be resifted prior to use because contents settle during shipping and storage.

Place a sheet of parchment paper or wax paper on a work surface. Fill a 1 cup measure with flour. Tap the flour down.

1. How much does the cup of tapped-down flour weigh?

Add additional flour to the cup to bring the tapped-down flour to a full cup measure.

2. How much does the cup of flour weigh with the additional flour?

3. What is the difference between the weight of the original cup of flour and the weight after the additional flour was added?

Pour the flour out of the cup onto the parchment paper or wax paper. Stir the flour with a spoon to add volume. Gently spoon the stirred flour into the 1 cup measure (do not tap or press the flour into the cup). Level the cup.

4. How much does the cup of stirred flour weigh?

5. How much does the remaining flour weigh?

Pour all of the flour in the cup back onto the parchment or wax paper and then sift the flour. Gently spoon the sifted flour into the 1 cup measure. Level the cup.

6. How much does the cup of sifted flour weigh?

7. How much does the remaining flour weigh?

8. What other ingredients are measured like flour?

9. How does sifting flour affect a recipe for homemade biscuits or cookies?

10. How does sifting flour affect a recipe for pancakes?

Activity: Researching Bread Recipes

Yeast bread is a versatile staple that can be served at any meal. There are three types of yeast doughs: lean dough, rich dough, and rolled-in dough.

Research a lean dough, rich dough, or rolled-in dough bread recipe. Respond to the following.

1. Prepare a report that identifies the function of each ingredient in the chosen recipe. Explain how the ingredients work together to produce the finished bread.

Prepare the chosen recipe according to directions. Serve the finished product to the class. Discuss the results of the recipe. Answer the following questions.

2. Are there elements of the finished product that could be improved?

3. What ingredients or procedures could be changed to produce a better result?

Activity: Mixing Quick Breads

Quick breads are made using one of three different mixing methods. These methods include the biscuit method, the muffin method, and the creaming method.

Prepare the biscuits recipe from Chapter 21: Baking and Pastry Fundamentals. Respond to the following.

1. Describe the texture of the biscuits prepared as directed.

2. Describe the flavor of the biscuits prepared as directed.

Prepare the biscuits recipe from Chapter 21: Baking and Pastry Fundamentals, substituting room temperature butter for the cold butter. Respond to the following.

3. Describe the texture of the biscuits prepared with room temperature butter.

4. Describe the flavor of the biscuits prepared with room temperature butter.

5. Compare and contrast the biscuits prepared as directed and the biscuits prepared with room temperature butter.

Prepare the blueberry muffins recipe from Chapter 21: Baking and Pastry Fundamentals, making sure to mix the batter only until moistened. Prepare the recipe again, mixing until the batter is smooth. Respond to the following.

6. Compare and contrast the interior appearance of the baked muffins.

Activity: Analyzing Cake Preparation

Before mixing cake batter, all of the ingredients should be at room temperature and each ingredient should be weighed separately for maximum accuracy.

Identify each of the following as the two-stage, creaming, chiffon, or sponge mixing method.

1. Eggs and sugar are warmed and whipped to create volume and incorporate air before any other ingredients are added.

2. Fat and sugar are mixed until fluffy, eggs are then added, followed by alternately adding dry and liquid ingredients to the fat, sugar, and egg mixture.

3. All dry ingredients, fat, and part of the milk are blended at low speed while the egg and remaining milk are mixed separately and added to the batter in thirds.

4. Whipped egg whites are folded into a batter made from flour, egg yolks, and fat.

Respond to the following.

5. Why is the oven set at the lowest temperature possible in stage one of the baking process?

6. Describe the appearance of a cake during stage two of the baking process.

7. Describe stage three of the baking process.

8. Describe stage four of the baking process.

9. What will happen to a cake prepared in an oven with the temperature set too low?

10. What will happen to a cake prepared in an oven with the temperature set too high?

11. Describe how to test a cake for doneness.

Activity: Piping Icing

In order to create decorative patterns and shapes with icing, a pastry tip is inserted into the pastry bag before it is filled. Many different pastry tips are used to make various designs.

Obtain the following pastry tips: round, open star, closed star, basket weave, leaf, petal, and ruffle. Fill a pastry bag with prepared icing and perform the following. Have an instructor check the result and sign below when complete.

1. Pipe straight lines, squiggles, and printed letters using a round pastry tip.

Piping Straight Lines Proficiency

Instructor:____ Date:_________________

Piping Squiggles Proficiency

Instructor:____ Date:_________________

Piping Printed Letters Proficiency

Instructor:____ Date:_________________

2. Pipe squiggles, scallops, and stars using an open star pastry tip.

Piping Squiggles Proficiency

Instructor:____ Date:_________________

Piping Scallops Proficiency

Instructor:____ Date:_________________

Piping Stars Proficiency

Instructor:____ Date:_________________

3. Pipe scallops and stars using a closed star pastry tip.

Piping Scallops Proficiency

Instructor:______________________________ Date:____________

Piping Stars Proficiency

Instructor:______________________________ Date:____________

4. Pipe a basket weave pattern, squiggles, and scallops using a basket weave pastry tip.

Piping Basket Weave Pattern Proficiency

Instructor:______________________________ Date:____________

Piping Squiggles Proficiency

Instructor:______________________________ Date:____________

Piping Scallops Proficiency

Instructor:______________________________ Date:____________

5. Pipe a ruffled leaf and a smooth leaf using a leaf pastry tip.

Piping a Ruffled Leaf Proficiency

Instructor:______________________________ Date:____________

Piping a Smooth Leaf Proficiency

Instructor:______________________________ Date:____________

6. Pipe a tapered petal, round petal, and tapered round petal using a petal pastry tip.

Piping a Tapered Petal Proficiency

Instructor:____ Date:_______________

Piping a Round Petal Proficiency

Instructor:____ Date:_______________

Piping a Tapered Round Petal Proficiency

Instructor:____ Date:_______________

7. Pipe squiggles and scallops using a ruffle pastry tip

Piping Squiggles Proficiency

Instructor:____ Date:_______________

Piping Scallops Proficiency

Instructor:____ Date:_______________

Activity: Preparing Pies and Pastries

Both the proper proportion of ingredients and proper mixing techniques are essential to preparing successful pies and pastries.

Prepare a flaky pie dough. Roll out and blind bake the pie dough. Respond to the following.

1. Describe the texture of the piecrust.

2. Describe the flavor of the piecrust.

Prepare a flaky pie dough, substituting butter for shortening. Roll out and blind bake the pie dough. Respond to the following.

3. Describe the texture of the piecrust prepared with butter.

4. Describe the flavor of the piecrust prepared with butter.

5. Compare and contrast the piecrust prepared with shortening and the piecrust prepared with butter.

6. What are the advantages and disadvantages of using shortening in a piecrust?

7. What are the advantages and disadvantages of using butter in a piecrust?

Prepare the meringue recipe from Chapter 21: Baking and Pastry Fundamentals. Respond to the following.

8. Describe the texture of the meringue.

9. Describe the flavor of the meringue.

Prepare the meringue recipe from Chapter 21: Baking and Pastry Fundamentals, increasing the sugar to 32 oz to produce a hard meringue. Respond to the following.

10. Describe the texture of the meringue prepared with more sugar.

11. Describe the flavor of the meringue prepared with more sugar.

12. Compare and contrast the hard meringue and the soft meringue.

13. Identify a possible use for a hard meringue.

14. Identify a possible use for a soft meringue.